Contemporary Problems in Geography

The general editor of *Contemporary Problems in Geography* is Dr. William Birch, who is Director of the Bristol Polytechnic. He was formerly on the staff of the University of Bristol and the Graduate School of Geography at Clark University in the USA and he has been Chairman of the Department of Geography in the University of Toronto and Professor of Geography at the University of Leeds. He was President of the Institute of British Geographers for 1976.

Alan Wilson is Professor of Regional and Urban Geography at the University of Leeds. After reading mathematics at Cambridge he has served as Scientific Officer at the National Institute of Research in Nuclear Science, Research Officer at the Institute of Economics and Statistics, University of Oxford, Mathematical Adviser at the Ministry of Transport, and Assistant Director of the Centre for Environmental Studies. His publications include *Entropy in Urban and Regional Modelling*, *Papers in Urban and Regional Analysis*, and *Urban and Regional Models in Geography and Planning*.

Michael Kirkby is Professor of Physical Geography at the University of Leeds. He has done research at Cambridge and at the Johns Hopkins University and previously taught at the University of Bristol. In 1972 he published *Hillslope Form and Process* with M.A. Carson.

Soil and Vegetation Systems

SECOND EDITION

Stephen T. Trudgill

CLARENDON PRESS·OXFORD
1988

Oxford University Press, Walton Street, Oxford OX2 6DP

Oxford New York Toronto
Delhi Bombay Calcutta Madras Karachi
Petaling Jaya Singapore Hong Kong Tokyo
Nairobi Dar es Salaam Cape Town
Melbourne Auckland

and associated companies in
Beirut Berlin Ibadan Nicosia

Oxford is a trade mark of Oxford University Press

Published in the United States
by Oxford University Press, New York

First edition published 1977
Second edition published 1988

British Library Cataloguing in Publication Data

Trudgill, S.T.
Soil and vegetation systems.——2nd ed.——(Contemporary problems in geography).
1. Soils 2. Plant-soil relationships
I. Title II. Series
631.4'1   S596.7

ISBN 0-19-874139-1
ISBN 0-19-874138-3 Pbk

Library of Congress Cataloging in Publication Data

Trudgill, Stephen T. (Stephen Thomas), 1947–
Soil and vegetation systems.

(Contemporary problems in geography)
Bibliography: p.
Includes indexes.
1. Plant-soil relationships. 2. Mineral cycle (Biogeochemistry) 3. System
analysis. I. Title. II. Series
S596.7.T78 1988   582'.0522   87-24010

ISBN 0-19-874139-1
ISBN 0-19-874138-3 (pbk.)

Set by Colset Pte Ltd, Singapore
Printed in Great Britain
at the University Printing House, Oxford
by David Stanford
Printer to the University

. . . how great is the difference between what the man on the road sees and the man in the field experiences. The man on the road does not see the immediate reality: [yet] he does see something which they in the field do not see, he knows something that they in the field do not know. He sees the whole. He may only see enough to call it picturesque. The artist is the man on the road with vision. He truly sees the whole, he perceives the Divine Harmony. His task is to reveal the whole to those who are submerged in the part, to unveil the harmony which is really on earth, and thus lessen the burden of life.

JOHN STEWART COLLIS, *The Worm Forgives the Plough* (1973)

Natural science does not consist of ratifying what others have said, but in seeking the causes of phenomena.

ALBERTUS MAGNUS, ?1193–1280

There are two kinds of biologists, those who are looking to see if there is one thing that can be understood and those who keep saying it is very complicated and that nothing can be understood—you must study the *simplest* system you think has the properties you are interested in.

CY LEVINTHAL, quoted in J. R. PLATT, 'Strong Inference', *Science* (1964), 146

Preface to the Second Edition

This book has been revised but the central theme remains the study of inputs, outputs, reactions, and stability in soil and vegetation systems. The basic interest is in the interactions of components of systems. We still tend to teach aspects of soils, hydrology, vegetation, and weathering as separate subjects, yet it is even more clear today that the environment can react as a whole when stressed at a particular point. Thus, I have found that this book has been considered useful in those educational institutions where an understanding of the holistic nature of ecosystems is seen as important. In other institutions, I've formed the impression that the book has not been adopted because it does not fit into subject divisions enough, with weathering still being taught in geomorphology, atmospheric and hydrological processes in climatology, and hydrology and cycling in biogeography. However, the intention remains to go beyond these divisions and to look at system components and their relationships. Many people have said that they have found it a useful stimulus for seminar programmes and I am thus pleased that the book has not only been seen as a text book but also as a basis for discussion.

In this Second Edition I have not updated the book to include as many recent studies as possible because many of them have been ones which have calibrated known relationships: conceptual advances and changes in awareness have been far fewer than the burgeoning wealth of individual process studies. I have often felt that field and laboratory studies either quantify the obvious or hint at the unobtainable—saying how complex things are—but every little while someone carves off a bit of the unobtainable and makes it obvious; I have thus tried to incorporate these sorts of advances in this new edition. I have also studiously avoiding responding to those who reviewed the first edition, one of whom concluded that the book was 'turgid', and another 'in no way turgid'; others stressed that there was 'an almost perfect balance of theory and illustration', or that the worst aspect was 'the lack of balance between theory and example'.

As well as the entertainment provided by book reviewers, I would like to thank all those others who have given me criticism and encouragement since the First Edition and those who have recently increased my awareness of issues and processes. Notably, Wayne T. Swank of Coweeta Hydrologic Laboratory, North Carolina, USA; Staff of the Institute of Ecology, Athens, Georgia; Staff of the Department of Environmental Sciences, University of Virginia, Charlottesville, USA and also of many Universities and Institutes in Australia and New Zealand, notably the Centre for Resources and Environmental Studies, ANU, Canberra.

New friends and colleagues continue to provide invaluable support, especially Liz Cole, Nigel Coles, Tim Mitcham, Tim Burt, David Job, Keith Chell, Dave Roper, Helen Springall, Annie, Simon, Ainé, Jim Hansom, David Thomas, Steve Plummer, Steve Maddock, and Fang. Neither should I forget the Sheffield student, who shall be nameless, who was caught cheating in exams: it transpired that he had copied out a chapter from the First Edition of this book, word for word, and smuggled it into the exams. I unwittingly marked the paper before I realized what had happened, giving myself a low 2.2, commenting that there were some good ideas, but badly expressed! Having been thus shamed, I have endeavoured to improve the text accordingly in this new edition. I hope that I will now earn at least a low 2.1.

STEPHEN TRUDGILL

Sheffield, January 1987

Preface (1977)

Having attempted to write a reasonably detailed book about environmental processes I find it difficult not to agree with Spike Milligan who wrote 'This damn book nearly drove me mad' (in his foreword to *Puckoon*). The major challenge in this kind of topic is the desire to be accurate about specific points together with the necessity of making bold generalizations about the whole system. What is to be seen in the following pages is a compromise between the two. The maddening thing is that once it is realized that the environment is a mutually reactive system, it becomes extremely difficult to isolate and discuss specialized points in a simple way without at the same time being tempted to discuss every interaction and interrelationship which exists in the system. Accordingly, the solution which has been adopted here is to deal with each component of soil and vegetation systems in turn and to build up a sequential picture of the whole system towards the end of the book. Originally, it was intended to deal in detail with the three main systems of flow and cycling: nutrients, energy, and water. However, this proved to be too daunting a task and the focus of the book is the consideration of nutrient elements in soil and vegetation systems.

The first section considers the general problems inherent in attempting to understand soil and vegetation systems. The discussion focuses on some of the ways in which the problems may be tackled by model building especially the building, of systems models.

In the second section, Chapters 2 to 6 deal with nutrient systems. The nutrients which are common to weathering processes, atmospheric input, leaching, and cycling are discussed here. In order to limit the discussion and to simplify the processes considered, elements such as nitrogen, carbon, and phosphorus are not dealt with in detail; elements such as calcium, magnesium, potassium, and silicon form the basis of the discussion. While such a limitation is regrettable, the picture that emerges is complex enough without further addition. The basic nutrient systems model is described in Chapter 2 and then each input and output component is discussed in Chapters 3 to 6.

Lastly, in the third section, soil and vegetation systems are viewed as whole, functional entities. In Chapter 7 the components outlined in Chapters 2 to 6 are discussed in a detailed nutrient model. Variations of this model are then considered for examples of specific environments and for examples of specific elements. The scope of the book is widened and the question of stability and change is discussed in Chapter 8 in terms of general models and then of disturbance and recovery in the face of man's actions.

The book is intended for the 'middle ground' higher education reader. It assumes a certain basic, elementary knowledge of soil, vegetation, weathering, and hydrology. At many points, however, reference is made to more detailed texts and research papers for more advanced study. In the context of British higher education, it is intended as a second-year text. However, some of the more basic, outline models will be useful at first-year level while some of the discussion obviously leads into third-year courses. The coverage of topics in this book is uneven but it is not wholly intended as a factual reference text. Reference is made to such texts, as appropriate, throughout. The book is concerned more with ideas and concepts. It is hoped that the reader will be stimulated in terms of ways of thinking and in terms of approaches to environmental problems. In many ways, it is intended as a step back from detail and a pause for reflection about the nature of soil and vegetation systems. Interdisciplinary study and transdisciplinary study are often identified as worthy causes; I have tried to identify the problems and potentials of such approaches in a part of the environmental system. If I have done nothing else but highlighted the difficulties and merits inherent in such approaches, I shall feel that I have succeeded.

I am indebted to a large number of people who have helped me in numerous ways during the course of writing this book. Mike Kirkby has patiently been through more drafts than I care to remember and made a large number of useful suggestions. Dave Briggs has helped in many ways and has made many noble attempts at improving my style and grammar. Other people have helped in various ways. They include Ian Laidlaw, Frank Courtney, Dingle Smith, Rick Cryer, Len Curtis, Keith Smith, Steve Nortcliff, Pete Smart, Tim Atkinson, Charles Curtis, Mike Waylen, Ian Statham, Ann Briggs, Chris and Eve Gilmore, Laurence McCulloch, Maggie Calloway, Tony Thomas, and many other people too numerous to mention. Thanks are due to the professional and other staff in the Geography Department at the University of Sheffield for providing a friendly working atmosphere, and special thanks must go to Alan Hay who has the misfortune to occupy the room next to mine in the Department. As such he has borne the brunt of my endless quests as to how to spell and how to best describe and explain things. I would also like to thank the staff of the Botany Department of the University of Sheffield and the students and staff of the Field Studies Council courses who have provided friendly discussion and help at various times. I am grateful to the secretarial staff of the Geography Department for struggling with my atrocious manuscripts and rendering them into impeccable typescript and also to the cartographic staff for help with the diagrams. I would especially like to thank those people who fostered my early interest in the natural environment, especially members of the Norfolk and Norwich Naturalists' Society and Janet Smith, the local librarian, who did so much for me in the way of appeasing my awakening avid appetite for natural history books. Lastly, I would like to thank my parents who have always supported and encouraged me and my interests.

STEPHEN TRUDGILL

Sheffield, September 1976

Contents

Contents

I
Basic Approaches

1 Systems Modelling of Soil and Vegetation

Science never proves anything, it makes guesses and goes by them as long as they work well.

STEINBECK

1.1. Introduction

In the scientific investigation of our environment we are basically trying to understand the environment and the organisms that live in it. Our level of understanding is important because, not only is knowledge valued for the way it satisfies our intellectual curiosity concerning the world about us, it also forms the basis for creating our environmental management policies. Social, economic, and political factors usually determine the actual implementation of management plans, but it is clear that if our understanding of environmental systems is not adequate, then neither will our management plans be adequate. Quite often our increased understanding of environmental systems can come from the actual implementation of management plans, that is, when the results of implementation are different from those intended: side effects and unforeseen reactions then tell us something more than we previously knew about processes and interrelationships. However, the logical basis for future management stems from the prior scientific investigation of the environment simply because it is better to foresee and avoid problems rather than to improve knowledge at the expense of possible damage to the environment and its resources. In addition, while it is clear that management relates to understanding, it is also clear that the type of understanding relates to the type of scientific investigation. This is especially true in terms of the way we approach scientific investigations in relation to our perceptions of the environment.

Traditionally, we see environmental systems as complex phenomena, which many factors operating in complicated, multi-faceted interrelationships. Our traditional reaction to this is either to be dissective, and to break down environmental processes into small, defined compartments in a specialist manner, or to look at systems in a broad, generalist manner. Both these approaches deal with perceived complexity, but in different ways; the first focuses on selected items and ignores other specialist details, the second looks only at the broad, main relationships and ignores subsidiary details.

However, there is a tension between generalization and specialization. Clearly, if the problem which the research worker is to tackle is to be of manageable size, it has to be defined and limited in some way. On the one hand, if the topic is a specialized one, closely defined and carefully limited, the work may be on the relationship between, say, soil nutrient levels and forest productivity. On the other hand, if the topic is a more general one, and is concerned with the ways in which several factors interrelate, the work may then include a study of the ways in which a number of factors influence forest productivity, such as soil nutrient levels, water status, soil structure, soil organic matter, soil depth, and so on. The limitation of the first approach is that while it may be possible to work in some detail on the relationships, such work would be only part of the story, and many other factors also influence crop productivity. Thus it might appear that the second approach would be better. The limitation, however, of the second approach is that while the inclusion of many more factors makes the study more realistic, this can also make the study more unwieldy. The research project would then either have to be a large one, with many resources and considerable manpower or it would tend to be superficial, not treating each factor in any detail.

One common resolution of these problems is clearly the factorial experiment where only the factors of interest are varied and all other factors are kept as constant as possible. Thus, for example, using subsamples of identical growth media, and under identical environmental conditions, the effects of nutrients on plant productivity can be studied in control situations. This is a very reasonable way to proceed for it can provide fundamental theories about relationships which may be applied to a wide variety of situations. However, in nature, we perceive that many factors operate together in a way that is different from how they operate in isolation under control conditions. But, if more realistic experiments are undertaken, with many factors involved, the results may tend to be more specific to particular combinations of factors which may then only apply to local situations.

We could thus feel that a specialist approach is too limited in its factors, but the work might apply, with calibration, to many possible situations. Alternatively if the work is general and multifactorial, we might feel that it is superficial and it may also apply to a few situations.

Do we necessarily need to follow these kinds of approaches? It is, perhaps, our perception of complexity which guides our thinking because we have a tendency to be dissective and then to think of aggregating the dissected parts into broader systems. An alternative is to be more *holistic* and to look at the whole system both in general terms and also, more importantly, in the ways in which the specialist details fit into the overall scheme. This thus reduces the tension between specialization and generalization since we can see both how the individual parts relate to each other and how they relate to the whole; by the same token we can also make our view of the overall system less superficial by the understanding of how the overall system is comprised of detailed, specialist components. Such a holistic philosophy is central to systems thinking because it tends to stress the study of the relationships between the the individual components themselves and between the components and overall system.

A key point to remember, however, is that nature does not know our boundaries. The world of components, factors and systems interrelationships is our world. The human brain is one which does not receive all information equally, but it filters and selects in terms of preceived significance according to perconceived concepts. We perceive complexity and thus we deal with this by selection. What we then have to remember is that our selective, partial view of the world is not the real world. The natural environment has elements of mutualism and sympathetic interaction which even holistic systems models may fail to allow for and specify. However, if we bear in mind such limitation, holistic systems models can represent a fundamental and important way of thinking about the environment and about the likely consequences of our environmental management actions.

1.2. Modelling

Model building can be viewed as a sequential structuring of ideas concerning the workings of a system. It operates by the initial isolation of one or two simple components of the system under investigation followed by the study of their interrelationships. Once the significance of these relationships has been specified further attributes can progressively be built into the model until it achieves a level of explanatory power which is sufficient for the management of the system.

To illustrate this consider the use of a car. When someone gets into a car for the first few times in order to drive it they build up a conceptual model of how it works. They study the relationship between the clutch pedal and the gear lever; between the accelerator and speed; between the steering wheel and the direction of travel, and so on. They build up a model—or picture in their mind—of how the machine works in order to manage it successfully.

But the system can be managed at several different levels. One can build up a knowledge of clutch plates and gear cogs; of carburettors; of king pins and tracking rods, and so on. One then understands the system more fully and can manage it more successfully. The difference between the person who only knows where the petrol, oil, and water go and the person who can service and repair their own car is obvious. In the former case, the person only has a simple, fairly crude working model of the car. Internal details of the system are not known but the system may be operated adequately—but not always completely successfully. Detailed knowledge of the system is not in fact needed until the car breaks down. Then further knowledge has to be built into the model so as to ensure that the management of the system continues to be successful.

The process is similar with environmental model building, but the system is a rather larger and somewhat more complex one. Furthermore, it was not built by human agency and therefore no one person or group understands the over-all design. Scientific discovery and model building are aimed at perceiving designs in and functionings of environmental manipulations under management so that we may manage it successfully and minimize detrimental effects.

The environment is approachable in the same step-by-step analytical manner as the building-up of a model of a car. The researcher may select, for example, the relationship between plant growth and soil water content; then the relationship between soil water content and rainfall, and so on, in order to build up a *dynamic working model* of a soil and vegetation system. Clearly, the researcher with the deepest understanding of the system will be able to manage it more fully and more successfully—and will know what to do when things go wrong.

The success of management policies depends not only on the knowledge of the manager but also on whether or not that knowledge is actually used. The first question is thus whether or not the degree of knowledge possessed by the manipulator is sufficient to foresee and avoid problems and side effects. It is when the limits of knowledge are overstepped that an adverse effect may occur, such as the erosion or acidification of soil. The second question, however, is not simply whether something is known, but whether it is, in fact, applied. Quite often, adverse effects are seen even though somebody, somewhere knew how to avoid them. Knowledge is always used—or not used—within political, economic, and social frameworks where there are limitations of communication and often a selectivity of purpose. Thus adverse environmental effects can be seen either because it was not known what the outcome was going to be or because it was known but the knowledge was not applied. It is often the case that an adverse effect promotes the revision of models of basic knowledge or, alternatively, it may promote the application of existing knowledge. Thus, society has often to be motivated by an awareness of a problem in order to apply the relevant knowledge or to promote the relevant research. We have to remember that we live in political systems where there is more kudos to be gained from the demonstrable solution of disaster than from the avoidance of a possible problem, which is always less tangible.

Whatever the nature of the society, the success of environmental management actions is clearly controlled by the limits of the conceptual models of the environment which the society holds and uses. If the actions of the society are seen as unsuccessful, then further existing models will have to be brought into play or knowledge will have to be revised and added to. The question arises, however, of whether it is really necessary that a disaster should be the first evidence the society has which shows that its environmental models are inadequate. Clearly, a driver need not crash a car in order to learn how not to drive it.

One of the ideas of model building is that a process of trial and error can be carried out in the laboratory, the computer, and the mind rather than in the actual environment, where we run the risk of damaging our resources. Moreover, the objective of *resource management* is to conserve our resources so that the options on their uses may be kept open for the future. The point of model building, especially that which involves simulation of the real world by scale models or mathematical formulation, is that it does not experiment with actual resources until the model is fairly well developed. Applied in this way, an approach which involves modelling helps to conserve the resources for the future. Naturally, the testing of large-scale models based on the operation of new ideas always runs the risk of disaster and side-effects. But the exploration of the environment and the discovery of how it works can be made one step safer by the introduction of model building because it may be possible to foresee some disasters, side-effects, and also the benefits before management actions are taken, and then to manipulate our resources for optimum advantage. Moreover, the concept of monitoring is important here. Models may indicate what the key operators or the most sensitive factors of a system are, and these can be monitored just as engine function can be monitored by an oil or temperature gauge in a car. Monitoring can thus be used to forecast potential danger before damage occurs. It is relatively easy to look at the last disaster and to legislate so that it may not happen again. This may be termed *retrospective* planning. Using predictive model building it may be possible to foresee a possible disaster and try to prevent it. Predictive model building is not so much a matter of what will *necessarily happen* but what *may* happen *if* a current trend is allowed to continue or if some interaction is caused or is allowed to happen. The research worker often begins by trying to model purely natural systems, at least in principle, and then moves on to assess the role which mankind has in altering these systems. As a result, the worker hopes to be in a position to assess whether man plays a beneficial or a detrimental role in the system and to attempt to maximize the benefits. Value-judgements of benefit and detriment may, however, be difficult in this area for, obviously, there are different groups of mankind with differing levels of awareness and aims. In addition, short-term benefits may bring long-term detriments. Discussions of what constitutes a benefit and a detriment are clearly difficult. But at least an approach to the solution of environmental problems which involves model building, environmental testing, and monitoring holds out hope for the solution of these problems on a rational basis.

Model building thus seeks to build up a series of relationships in order to achieve sufficient explanatory power for a particular purpose. As has been mentioned, the operation of model building may follow the path of questioning such as: 'What if so and so happened, or was done, what would be the result?' So a model—be it conceptual, hardware, or mathematical—may be built by studying the outcome of different combinations of possibilities. The operation of the model may be studied both with and without the addition of

specified factors. Subsequent to the definition of the problem (i.e. specification of what knowledge is required) then the process can be summarized in
the following steps:

1. Seek simple relationships between a few variables thought to be of
    importance.
2. Define the outcome of these relationships.
3. Test the outcome against the real world.
4. Define the predictive value and limitations of the model.
5. If the model is valid for a particular purpose, put it into operation; if
    the limitations are too great for that purpose then build in further
    variables or modify the model in some other way.
6. If modified, redefine the outcome and test again.
7. Recycle this procedure until the model has sufficient explanatory
    power for the purpose in question.

This process can never really stop—with a final solution—for the purpose is
likely to be redefined, new knowledge is continually being gained, and
unforeseen side-effects and problems invariably occur. Thus the modelling
process can be illustrated in the form of a continuous flow diagram with
recycling loops (Fig. 1.1). This whole operation is a very large one, involving
the context in which the contents of this book are placed. Parts I and II are
directly concerned with steps 1 and 2. Steps 3 and 4 are involved in the latter
part of the book (Part III).

A problem with building models of soil and vegetation systems is the
complexity of environmental systems. As has been stressed earlier, it is
necessary to use some simplifying assumptions in order to begin with a model
of manageable proportions. But the problem of devising any model is that, as
model building proceeds, the researcher tends to build in more and more
factors. While the model may thus achieve greater explanatory power it may
also become unwieldy. The point is, of course, that ultimately so many
variables could be built in that the model will be so massive and complex as to
be as incomprehensible and as difficult to manage as the real world itself.
While one of the aims of model building is to replicate the real world, the
human mind can only deal with relatively small units of information at any
one time. Furthermore, it needs clear, decisive instructions that can be put
into operation in management. Thus there is a need both for simplicity and
for the accurate representation of a complex system.

The difficulty lies in meeting these two requirements. The real world is very
complex and many things affect many other things: one action rarely has one,
single consequence. For example, consider the following statements:

The natural environment is itself an integrated system, a complex web, if stressed at a
specific point, usually responds as a whole. (Commoner, 1968.)

The ecosystem is an indivisible complex in which no animal or plant exists in total

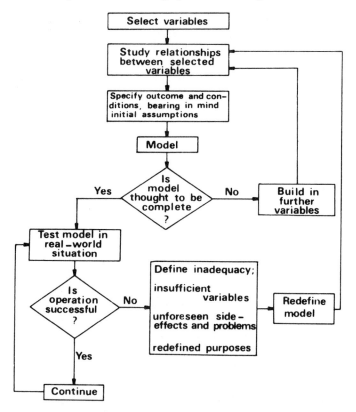

FIG. 1.1.   The process of model building and testing

isolation and no factor, physical or biotic, operates in complete independence. (Glen, 1967.)

However, it is necessary to analyse the complex world into single items, or sets of related items, and to rationalize it into principles concerning the behaviour of the whole assemblage in order to be able to deal with it. We need to divide the 'indivisible complex' in order to comprehend it.

A key concept involved in trying to resolve this paradox is that everything is related to everything else *to a greater or lesser degree.* Thus environmental analysis attempts to identify *the most important relationships* which exist in an environmental system. It also seeks to identify the variables which control the relationships involved. The aim is to produce models of soil and vegetation systems where the key variables are differentiated from the incidental or noise variables and where the important relationships are identified. The points which have been made by Runge (1973) are pertinent to this discussion and can be usefully presented here in an abbreviated form:

Soils are complex systems to study because of the many potential variables that influence soil variables. However, some variables are more important in controlling soil development than others. If this simplification of the extraction of key variables can be made then our chances of discovering a portion of the real world in a manageable research project are increased. This simplification of the soil system is referred to as modelling. One must be sure to remember that the model is not the truth itself but is only of some use if it leads to a more rapid discovery of the truth. (After Runge, 1973, p. 183.)

Variables such as soil moisture content and nutrient availability can isolated as key variables in terms of a plant productivity system. Other variables of lesser significance can subsequently be ranked in order of importance. It can then be evaluated which variable is likely to be most sensitive to manipulation and which is likely to lead to the greatest plant response. For example, soil nitrogen content has been identified as a key variable in plant productivity (Cooke, 1982, p. 3). Thus plant growth will be highly sensitive to the manipulation of this variable.

However, a model of plant productivity based solely upon soil nitrogen content would be inadequate to raise plant productivity to a maximum. Factors such as water supply and soil aeration are also important. The point is that where other variables are adequately catered for, nitrogen tends to become the dominant control. If nitrogen supply is adequate, then the manipulation of water supply may become the dominant control. The situation is therefore flexible. Nitrogen, water, and aeration, amongst other factors, all have their effect; their order of importance depends upon the overall situation according to *Liebig's Law of the Minimum*: the factors at the minimum level tends to control the whole system. Environmental systems are characteristically multifactorial and thus it is necessary to discover the *mutual effects* of several variables acting at once. Moreover, it is important to specify how the variables may change over time and how they may mutually adjust to each other as one or more alter.

Model building at this stage begins to become difficult and complicated unless some unifying and simplifying concepts are used. By using those concepts provided by systems thinking it is possible to isolate and study the relationships of selected variables in a multifactorial system. In general terms systems thinking, through its concept of interacting components, enables the research worker both to divide the indivisible complex and to remain aware of the integration that exists in environmental systems. In other words, one can not only break down the system into parts which are understandable, but also put them together again and show how they interact as a whole. We must, of course, bear in mind that while we may invent boundaries in order to grasp an understanding of nature we must remove them at a later stage in order to have a valid understanding of how nature in fact works. It is necessary to think of the child who takes a clock apart so that he may understand the parts—but who is lacking in his ability to reassemble those parts into a working whole.

1.3. System definition and description

Systems thinking is a body of ideas concerned with the state of matter and the factors that influence that state. It is concerned with the organization of matter and the dynamics of the processes that lead to that organization.

Systems model building begins with the specification of a system boundary followed by the study of transfers of energy and matter across that boundary. It proceeds by the successive evaluation of key transfers which control changes of state within the boundary. It is thus concerned with relationships within the system and the relationships of that system with its surroundings. A systems model is a representation of the interaction of system components (and subcomponents), their integration, and their relationships with external variables.

A distinction can be made between *systems analysis* which, at its most specific, can involve the mathematization of organizational links within and between systems in order to optimize some particular attribute of the system, and *General Systems Theory* which proposes that all systems ultimately can be understood by the application of systems principles. Neither of these two is discussed in further detail here but reference may be made to Beishon and Peters (1972) for a discussion of the former and to von Bertalanffey (1962) and Harvey (1969, pp. 447–80) for discussions of the latter. This book is concerned more with the use of the operational concepts such as those originating from thermodynamics (Ramsey, 1971; Adkins, 1968) and those which have been translated into the environmental sciences with varying degrees of rigour by a number of authors (for example, in geology by Broecker and Oversby, 1971; Garrels and Christ, 1965; Berner, 1971; and as expanded for use in ecology by van Dyne, 1969; Watt, 1968; Richards, 1974; Dale, 1970; and in physical geography by Chorley and Kennedy, 1971). The concern here, in the context of soil and vegetation systems, is primarily with *Systems thinking*, that is with those concepts dealing with the dynamics of systems. The questions which are being asked are 'How does a system work in terms of inputs and outputs from other systems, in terms of linked functional relationships, and in terms of changes in the flows and storage of matter over time?' Systems thinking has been usefully reviewed by Emery (1969).

This type of approach can be of great use in the analysis of the environmental problems in that the reactions within systems may be modelled in terms of changes that occur in response to transfers across the system boundaries. This is helpful since, if manipulative management actions can be translated into terms of boundary transfers, the resulting reactions within the system may be specified or predicted.

The *system* is defined as that part of the environment which is under primary consideration. Thus the research worker may talk of soil systems or plant systems or soil–plant systems, depending upon the purposes of his study. Once the system is defined then the *surroundings* of the system are

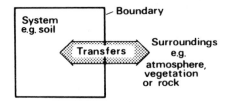

FIG. 1.2.   System definition

defined as being any other part of the environment not under primary consideration, but which it may be necessary to consider in order to understand the inner workings of the system (Fig. 1.2).

Some systems are more easily defined than others. The human body, for instance, can be easily defined as a system which is separate from the air surrounding it. A soil profile may be more difficult: some soils may be sharply differentiated from the underlying bedrock while others may merge very gradually into some unconsolidated deposit of sand and gravel. In the latter case some degree of arbitrariness may be involved in placing the boundaries of the system.

The emplacement of such a boundary may be made considerably easier if the research worker has a specific problem in mind rather than if he is trying to place a general purpose boundary. Thus a system boundary in a soil/parent material transition may be taken, for instance, as the lowest point in the soil profile where organic matter content falls below a certain minimal percentage. Above this, the material may be defined as part of the soil system and below this as the external parent material. This boundary can then be used to define the location for the study of transfers between the soil and parent material.

The boundaries of a system are thus placed in order to solve a problem. The key point is that system boundaries are not necessarily there to be discovered, some degree of artificiality is often involved and boundaries are placed by the researcher with a purpose in mind—but not, of course, without some basis selected from the real world. It is not to say that in reality boundaries exist of necessity, but rather that from reality boundaries can be inferred which appear significant to the researcher. This is an important point to bear in mind when comparing one system with another because some degree of subjectivity may have been involved in the initial definition of the system.

If it is so desired the system may be divided into a hierarchy of systems. Thus *supersystems, systems*, and *subsystems* may be described. Transfers between these component systems may be discussed as well as transfers between them and their surroundings. For example, a woodland ecosystem can be described as a supersystem, while separate soil and plant systems may be described at a second, lower level and root, soil pores, or clay particle subsystems be defined at a third level (Fig. 1.3).

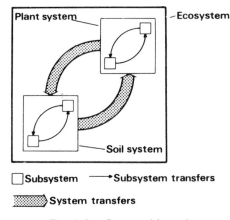

Fig. 1.3.   Systems hierarchy

The *state of the system* is specified by listing its *properties*, such as temperature, chemical composition, morphology, or other convenient or desired parameters. It is often possible to specify the state of simple systems very precisely. In these systems, for example, it may only be necessary to describe conditions of temperature and pressure in order to specify an attribute of state (such as volume) in a simple system. The more complex a system becomes the more properties or variables it may be necessary to describe in order to fully specify the system.

It has already been stressed that in complex environmental systems the number of variables involved may be legion. While key variables which largely specify the system may be sought, many different types of variables can be usefully indentified.

Key variables which control important relationships can be *external* or *internal* to the system. If they are external they are termed *independent* or *control variables*. For example, climate and lithology can be thought of as external control variables. Internally, for example, soil permeability can be thought of as a control of leaching. However, these internal key variables are not necessarily themselves independent variables. Soil permeability will, for example, be dependent upon soil texture, structure, and the formation of horizons. Thus there are several important things to consider:

1. The number of *variables* involved and the *relationships* between the variables.
2. Whether or not the variables are *external* or *internal* to the system under discussion.
3. Whether or not the variables can be viewed as *independent* or *dependent* of other variables under discussion.
4. Whether or not a variable is a *key* variable directing the over-all

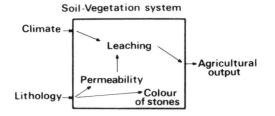

FIG. 1.4.   The relationships of some soil and vegetation system variables

system, or output, or whether it is just a *noise* variable which does not contribute fundamentally to the relationships under discussion.

Figure 1.4 shows the relationships of some soil-vegetation system variables. Variables considered are: climate, lithology, permeability of the soil, leaching, agricultural output, and colour of the stones in the soil. Defining the soil-vegetation system boundary as shown leaves climate, lithology, and agricultural output as external variables; the rest are internal. Climate and lithology are *independent* of the system. Agricultural output is a *dependent* variable of the system. Lithology is a control of soil permeability and climate is a *control* of leaching. Permeability is *dependent* upon lithology but is also a *control* of leaching. Leaching is *dependent* upon climate and is a *control* on agricultural output *via* its effect on soil nutrient status. Stone colour is *dependent* upon lithology, but as far as other variables are concerned it is just a *noise* variable.

Thus the relationships between internal variables may be very complex and often involve the operation of intervening variables. Therefore, for the purposes of study and comprehension, selection of just a few variables may be made initially, individual relationships being examined successively. Alternatively, the intervening internal variables can be ignored and the main relationships between the external input variables and external output variables can be studied.

Specification of state tends to go hand in hand with specification of changes over time. This is because, in environmental systems, energy and matter are continually flowing through the systems. Energy is continually being transferred from sunlight to plants and from leaf litter to soil organisms and then to predators and so on; water is continually flowing through the systems and chemical elements are also being transferred along with these flows. Environmental systems are therefore referred to as *open systems*, meaning that matter and energy are continually being transferred across their boundaries (Fig. 1.5). It is therefore necessary not only to specify the state of a system at any one time but also to specify any changes in its state through time. Thus, once the system boundaries have been defined and the transfers across its boundaries have been itemized, the modelling of temporal

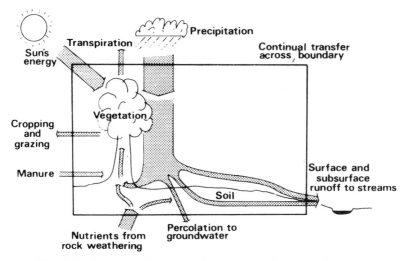

FIG. 1.5.   Inputs and outputs of an open environmental system

changes, stability, and equilibrium in the system become important operations.

Although much of the scientific research into soils, vegetation, hydrology, and solutes is specialist in detail and limited in scope, recent works on chemical weathering processes, nutrient cycling, and hydrological processes have often stressed the wider contexts of their results. This has made it feasible to attempt to discuss these topics in an interrelated sense. Thus, studies of weathering of rocks and soil minerals, and the concomitant release of plant nutrients, can be linked to those of plant nutrient uptake, nutrient cycling by vegetation, nutrient inputs from the atmosphere, and solute losses in drainage waters.

Once the basic processes have been specified (as discussed in Chapters 2–6 of this book), linkages can be evaluated and the balances between inputs, retention, and losses in whole soil and vegetation systems under different environmental conditions can then be discussed (Chapter 7). Chapter 8 then proceeds to discuss changes over time, and the stability of soil and vegetation systems, which are important considerations in environmental management.

From these discussions, it will be seen that scientific research can provide a fundamental basis for management, but, nevertheless scientific research does not adopt one, uniform method. Moreover, scientific method can often be dissective whereas the intention of systems modelling is to be holistic and integrative. It will therefore also be useful to conclude this Introduction by discussing briefly the nature of scientific method and the relationships between scientific method and systems modelling.

1.4. Scientific research, systems thinking, and environmental management

Explanations and descriptions of systems are necessary but not sufficient conditions for successful environmental management. Physical models derived from scientific research provide the basis for rational environmental management because it is easier and cheaper, in the long run, to work with nature than against it. However, there are tensions within scientific methodologies and between scientific research and environmental management which make it important to discuss both the internal processes by which science derives physical models and also the relationships between scientific research and the political, economic, and social frameworks in which environmental management actually operates.

There are two tensions which can be seen in the relationships between scientific research and environmental management. Firstly there is the tension within scientific methodology itself between the ostensible scientific methods described by those who write about scientific method and the actual practices carried out by scientists. Secondly, there is the tension between the type of science which is actually carried out and the needs of management in terms of a scientific basis for management plans.

In a discussive text, Haines-Young and Petch (1986) review the use of scientific method in physical geography, proposing that a deductive method of scientific enquiry is one of the few defensible forms of research. By a deductive method, they include the testing of ideas by observation and experiment under the umbrella of 'critical rationalism'. This is to say that there is a proposition about how the real world works, derived from previous work and/or intuition, which is tested using a crucial experimental design in order to see whether or not the proposition can be falsified. If it cannot be, then it is left open for further testing. Such an approach is supported by an important paper on scientific method by Platt (1964) who also stresses the importance of designing experiments to test ideas, and especially to falsify them. He writes: 'The man to put your money on is not the man who wants to make "a survey" or "a more detailed study" but the man with the notebook, the man with the alternative hypotheses and the crucial experiments' . . . and . . . "what experiments could *dis*prove your hypothesis?' Many scientists would pay lip-service to this approach. However, if 'good science' can be seen as the falsification of ideas about how the environment works by rigorous experimental design, this is not necessarily what all scientists do, nor is it always what practical managers need.

Many workers, rather than adopting a falsification procedure, in fact adopt a verification procedure, whereby evidence builds up to fit a theory, the conclusions often being that 'it appears to work like this . . .', rather than 'we have refuted this and this possibility, leaving others open for testing'. Many also freely admit that their work is non-experimental and is not undertaken in a problem-solving context; nor does discovery always come from the strict application of directed hypothesis testing, it often comes incidentally while working on something else. These points are admirably summarized by Rackham (1986) who, in an eminently scholarly book on the history of the English countryside, writes:

Historical ecology sometimes involves the set-piece methods of scientific research: problems defined in advance and information collected wherewith to solve them. But in many areas this will not work because the facts are too thinly scattered to justify deliberate search. Much of the material for this book has come my way in the course of other researches: facts have turned up and have been filed away until enough has been hoarded to suggest questions and answers. Insights may also come at random from travels made, or documents read, for some different purpose. I went to Texas to discuss Cretan archaeology, and what I saw made me revise my ideas on hedges.

Scientific methodology purists would tend to reject this point of view, although they might recognize that many people work in this way.

Haines-Young and Petch regard a verification approach as philosophically indefensible since no matter how many cases fit in with a proposed theory, there always remains the possibility of a case, as yet unfound, which will refute the theory. Despite this, there are innumerable papers in the scientific literature which have not involved falsification, perhaps because it is seen as irrelevant, too impractical or because the philosophical standpoint is simply ignored. Instead, research workers accumulate knowledge about how a system appears to work. This can be especially true of approaches to large systems, where the control of variables in replicated situations is often difficult to effect.

However, while it is often seen as difficult to use experimental design in the investigation of whole systems, it is not impossible. This has been shown by some large scale catchment experiments involving different treatments of replicate areas (e.g. Swank, 1986). Here, entire replicate catchments are set up on similar rock types and soils and under similar climatic conditions, but land use is varied, with, for example, plantations of coniferous trees or the retention of deciduous forests. Then different management strategies are applied, such as a variety of wholesale and selective felling strategies, a variety of road building strategies, and a variety of herbicide application strategies, and many others, all to determine the effects of different management treatments on water yield and on sediment and solute outputs. These large-scale experiments represent scientific method at its best because they involve effective control of management variables in realistic, whole system situations. They can therefore give the most useful results on how whole systems behave under management. However, they are costly and labour-intensive to set up. Thus if the resources of time, money, and land are not available for such large scale experimentation, the approach remains one of smaller scale, specialist research and the building up of ideas about whole-system processes in a verification sense and often it is this body of accumulated wisdom that practitioners use when drawing up management plans. Thus, with the exception of large-scale experimentations, management of large systems tends to be based on a 'rule of thumb' approach about how we think things work, rather than on the results of experimental tests, let alone a series of statements involving the falsification of ideas on processes.

This is an important point because often, in reality, practitioners are essentially drawing upon hypotheses which have not yet been refuted. A lack of experimental evidence or implementation testing can clearly be a source of management problems because there will always be the possibility that managers have got it wrong. The managers feel that there is sufficient evidence, in a verification sense, on which to base their actions, and this is supported by evident scientific approaches but these are often merely accumulated cases which fit in with particular theories. If, in fact, the actions are really based on as yet unrefuted hypotheses, then it can indeed be seen that there is great scope for mismanagement. Even if there is past experience of actions which have worked successfully in other places, they may not work in another. As Haines-Young and Petch suggest, verification is weak as there always remains the possibility of a case, as yet unfound, which will refute the theory. It can readily be seen, as proposed earlier in this chapter, that management actions on large systems are in many ways often the only tests of hypotheses about how such systems work, only providing evidence on whether or not our ideas are right once they are put into action. This pragmatic approach of accumulated evidence and trial and error in practice is pithily, if somewhat cynically, epitomized by the novelist Steinbeck (1960) when he observed that 'Science nevers proves anything, it makes guesses and goes by them as long as they work well'—as heralded at the start of this chapter.

If it is accepted that a defensible scientific method involves falsification, as promoted by Haines-Young and Petch, it is possible to observe that this approach, in fact, only tells us what we should *not* do, rather than providing guidelines for management: hence there can be a tension between scientific method and management. 'Good science' (falsification) is necessary to eliminate erroneous ideas, certainly, but this merely leaves options open on other ideas, if it is strictly adhered to. So on the one hand, some scientists are saying that falsification is the only defensible approach to science, while, on the other, such an approach does not necessarily produce positive guidelines for management: these tend to be based on verification and the accumulation of evidence. The problem with this is that there can then always be a case where the ideas do not work, the management action having provided a further test of the ideas and this, if we have got it wrong, can have deleterious consequences for the environment.

It can be argued, then, that there are problems with a rational approach to management based on the rigorous application of scientific method as is proposed by some advocates. In addition, management is not always placed wholly, if at all, in a rational context: political considerations often have a greater influence on management decisions than scientific knowledge, let alone the outcome of strict hypothesis testing. Frequently, simple descriptions of states are the only 'scientific' steps involved, together with value judgements on the desirability of the states. Here, the arguments become more ones of perception, aesthetics, and social pressures. This is not to say that scientific work on processes should not be important, but that frequently, in the real world, it may be ignored or take second place to more emotional arguments. If science is involved in any way, then it may be solely in terms of a descriptive survey of, say, the number of flowers or birds seen as desirable in an area to be affected by proposed management, in order to assess the value of an area. Thus, management plans may be implemented or quoshed irrespective of scientific knowledge about processes and system linkages, though such a knowledge is usually seen by scientists as a fundamental basis for rational management.

So, it can be argued that the need of managers is to have a body of accumulated, positive evidence which can be used as a basis in formulating their policies, which are then implemented within a societal value system, rather than for a series of tests which have falsified various options. Falsification is a crucial step in the negative sense of eliminating those courses of actions which will definitely have deleterious effects, but more is required: that is, positive statements of what should be done, what will work, and what the outcomes of various alternative actions could be.

We can conclude that many scientists work in a way which is parallel to the needs of managers by providing relevant evidence using a verification approach, but scientific methodologists often maintain that verification is indefensible. The problem with the more defensible falsification is that it may not provide a body of evidence which is practicable. Thus, the problem with basing management actions on accumulated bodies of evidence is that if such bodies can be seen as collections of as yet unrefuted hypotheses then this is clearly going to be a source of management problems in terms of unforeseen side-effects and problems.

In addition, a traditional, dissective, specialist approach to environmental components does not provide us with information about how whole systems might react under management actions. Clearly, what is needed is rigorous scientific research on whole systems, involving not only large-scale control experimentations on whole catchments and ecosystems, but also the modelling of whole systems in order to study the likely consequences of management actions. Such approaches act to reduce the tensions between scientific research, scientific methodologists, and management because false ideas can be eliminated by experiment and observation and the consequences of actions can be demonstrated and modelled by approaching whole systems in a scientific manner and in a way that is useful to managers. Even then, however, we are only dealing with probabilities: no management action can be 100 per cent successful because, quite apart from success depending on what the criteria for measuring success are, we are always operating in a state of partial knowledge.

The actual implementation of policies depends so much upon societal contexts, which vary markedly round the globe, and which are largely outside the scope of this book. What is seen as important is the way we arrive at our knowledge which can be, and should be, used as a basis for rational management. Discussing how whole systems behave is seen as an important focus because if we can predict reactions we can then be in a position to judge the desirability of future, managed states. Thus, taking these points discussed above, in this book whole systems are first discussed in outline (Chapter 2), individual components are then described (Chapters 2–6) and, finally, whole systems are discussed in more detail in the concluding Chapters (Chapters 7 and 8), especially in the ways they behave over time and their stability under management.

The judgement of a systems approach is thus seen in the way in which it can help us to manage our environment, both in terms of minimizing deleterious human impacts and maximizing the self-sustaining nature of environmental processes. Maybe we might even learn to let nature run itself, but assuming that we continue as we have started, by manipulating it for our own ends, it is to be hoped that we may realize that, in the long run, our ends and environmental well-being are the same thing. It is also to be hoped that we may understand our environment enough to be able to tune in more to environmental processes and, at the very least, realize that every environmental management action has wider implications than we thought and to consider these ramifications and implications. This point is at the very heart of systems thinking: 'Systems' can have a bad press because it appears to be a collection of box and arrow diagrams and a lot of jargon. This is often unfortunately the case (and this book is probably no exception)—but this is not the point of systems thinking. To my mind, if systems thinking only makes us stop and think about what the wider consequences of specific actions are for the environment as a whole, then it is a success.

II

Nutrient Systems: Components

2 Basic Models of Soil and Vegetation Nutrient Systems

Science seeks the unity under the chaos of natural phenomena.

BRONOWSKI

2.1. The basic components

There are five basic components of soil and vegetation nutrient systems:

1. Nutrient inputs from the weathering of minerals.
2. Nutrient inputs from the atmosphere.
3. Nutrient losses by leaching.
4. Nutrient cycling by plants and animals.
5. The store, in the soil, of nutrients which are readily available to plants.

The inputs and outputs interact with the nutrient store as shown in diagrammatic form in Fig. 2.1. The relationships illustrated are much simplified and in reality the interactions are far more complex. As discussed in Chapter 1, in order to deal with such a complicated system in a way which is comprehensible, it is useful to adopt two approaches. First, the interactions shown in Fig. 2.1 will be briefly described in broad outline in this chapter.

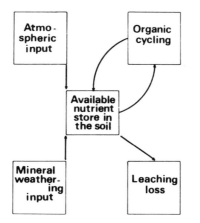

FIG. 2.1. A simplified model of the interaction of the five basic components of soil and vegetation nutrient systems

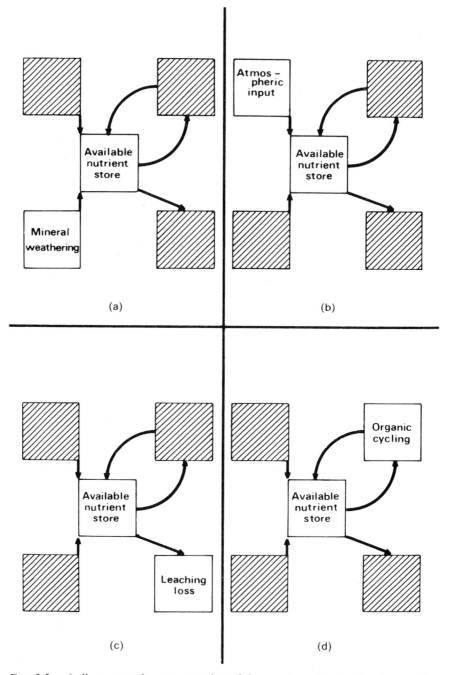

Fig. 2.2. A diagrammatic representation of the components of soil and vegetation nutrient systems to be discussed in Chapters 3 to 6

This will provide a framework for subsequent discussion. Secondly, in Chapters 3 to 6 each interaction will be isolated and studied in turn. The nutrient store will form the focus of attention in these chapters because each component interacts with it. In Chapter 3 the interaction between the mineral weathering input and the nutrient store will be discussed (Fig. 2.2(a)). Chapter 4 will deal with the atmospheric input (Fig. 2.2(b)). In Chapter 5 the relationship of the leaching output with the nutrient store will be examined (Fig. 2.2(c)) and Chapter 6 will look briefly at organic cycling (Fig. 2.2(d)).

While the understanding of a complex system is considerably aided by the use of the nutrient store as a focus of attention, the study of the interaction of each component with the nutrient store is, of course, artificial because it divides up a mutually reactive system. Therefore, although the interactions must be examined separately in order to be able to understand them, it is essential to discuss how the components interact within the context of the whole system. This over-all interaction will be considered in the section on whole systems, in Chapter 7.

The store of nutrients in the soil which is available to plants is an important constraint upon plant productivity and it is therefore important to focus attention upon it, especially in an agricultural context. Moreover, soil nutrient status has a significant effect upon soil properties and upon the quality of soil drainage waters. However, as will be discussed in Chapter 7, there is a great deal of mutual interaction in the system and plant productivity, soil properties, and soil water quality can all affect the over-all nutrient status of the soil system.

The major inputs of nutrients are from the weathering of minerals in soils and rocks and from the atmosphere. The latter may take the form of nutrients dissolved in rain-water or it may occur as a fallout of solid particles. The major output from the system is in mobile drainage waters (either in overland flow, soil throughflow, or groundwater percolation). This loss is usually in the form of a loss of solutes but, in the case of overland flow, the loss of sedimentary particles can also occur. The sedimentary particle loss may represent a loss of potentially weatherable minerals or, in the case of clays, cations which are adsorbed onto the particles may also be lost. Amorphous soil constituents, termed plasma (mostly organic and inorganic colloids), may be lost in throughflow waters. The plasma can be carried down slope through the soil matrix and again this represents losses of potentially available elements and of adsorbed cations.

Under natural conditions leaching losses can be minimized because of retention in the system of recycling. Elements taken up by plants are taken into the tissues. Leaves, stems, branches, and tree trunks are periodically returned to the ground surface and, on decay, release the nutrients which have been stored in the plant tissues. Also included in this process is the ingestion of plant material by animals and its return to the soil in faeces.

Under agricultural conditions plant and animal material will often be removed from the system by cropping. Nutrients which would have otherwise been returned to the soil by natural processes will be lost with the crop. Accordingly, the lost nutrients will have to be replaced by the addition of aritificial fertilizer or animal manure.

Nutrients added to the soil system, whether by weathering, in rainfall, from the decay of plant tissue and manure, or in fertilizers, will become part of the soil nutrient store. The store may be depleted by leaching losses or by plant uptake, the crucial difference being that (under natural conditions) the latter loss is made good by recycling. Cycling and leaching thus operate in opposite directions and it is often useful to think of soil nutrient status as a result of the balance between these two processes.

Thus, although the system is complicated and interrelated, it is clear that weathering inputs and nutrient cycling act to increase nutrient status while leaching losses and crop removal act in the opposite direction. The role of rainfall has to be carefully studied, however. In terms of amount, an increase in rainfall will act to promote nutrient losses in drainage waters but in terms of chemical characteristics it can also bring nutrients into the system. However, the role of rainfall in the nutrient credit and debit of the soil system is further complicated because it also brings substances to the soil which can be involved in weathering reactions. While these reactions will help to release nutrients for plant growth they can also release them for leaching. Matters of input, output, and retention are clearly of crucial importance to the availability of nutrients in the soil.

2.2. Nutrient availability in the soil store

Nutrients which are available to plants are retained in the soil in two forms. Firstly, cations are adsorbed onto clay-humus complexes and secondly, nutrients are present as solutes in soil water. The latter are not retained very firmly in the soil and can be lost readily by leaching but the former are far less prone to leaching and represent the principle store of exchangeable plant nutrients. Losses by leaching and uptake can be replaced by inputs from weathering, organic matter decay, and rainfall (Fig. 2.3).

The amounts of nutrients which are available for plant uptake at any one time can be viewed as a function of *inherent capacity for retention* and of *supply*. In terms of storage it is useful to distinguish between (1) the total nutrient storage capacity and (2) the extent to which this storage capacity is filled. The latter will depend upon the balance between supply and output. The former will depend upon such soil characteristics as humus content and clay content. Nutrient availability can thus be seen as involving two concepts—a relatively static one of capacity, determined by internal variables, and a more dynamic one of flow, determined by more external variables of input and output (Fig. 2.3). As discussed in Section 1.3 input will

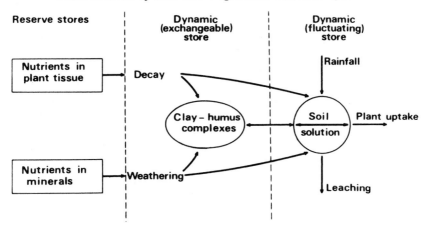

FIG. 2.3.   Losses, storage, and replacement of nutrients in the soil

be an external control variable while output will be dependent upon input minus retention. Retention involves two concepts—adsorption capacity, keeping nutrients in the soil, and water flow, as the freer the soil drainage then the more leaching losses are encouraged.

In many ways the availability of plant nutrients in different soils is analogous to the amounts of water which would be retained by different buckets which all leaked by different amounts and which were all being filled at different rates. Clearly, the largest buckets with the fewest leaks and the largest inflow will retain the most water; this is analogous to a soil with a high content of clay—humus complexes, with very little leaching (owing either to poor drainage or low rainfall or both) and high nutrient inputs from

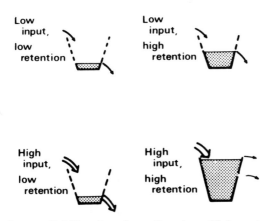

FIG. 2.4.   Nutrient availability viewed as a function of inherent capacity for retention and of supply: the analogy of leaking buckets

weathering or organic matter decay or from the atmosphere. Conversely, small, leaky buckets will contain little water, no matter how much water is poured into them. This is analogous to soils with a low content of clay-humus complexes and high leaching (owing to good drainage or high rainfall or both), irrespective of nutrient inputs. This analogy is illustrated in Fig. 2.4.

The shortcoming of this analogy is that it is valid only when the inputs are independent of the level of storage. This is not necessarily the case in soil and vegetation systems. In the soil nutrient system the removal of nutrient cations by plants and by leaching will act to encourage the release of further nutrient cations by the weathering of primary minerals.

The main components and interactions in the soil nutrient system have now been described and Chapters 3 to 6 will discuss each of these in detail.

3 The Weathering Input

It all depends what you mean by . . .

C. E. M. JOAD

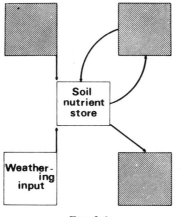

FIG. 3.1.

3.1. Introduction

In soils which have a dominantly mineral composition (i.e. excluding peats) under natural conditions chemical weathering is one of the most important sources of inorganic plant nutrients. In this context the concern is with the elements calcium, magnesium, potassium, sodium, silicon, iron, aluminium, and phosphorous. These are of major significance in plant nutrition. Other elements may be released by chemical weathering and, although they are required in smaller quantities, they are no less vital. These are sulphur, manganese, copper, zinc, molybdenum, boron, chlorine, and cobalt. In the discussion which follows the elements carbon, oxygen, and nitrogen are not considered as they are primarily involved in organic cycling.

The main concern of this chapter is with the chemical weathering of mineral particles in the soil (Jackson and Sherman, 1953). The interest lies in the release of chemical elements to the store of available nutrients. These available nutrients may be absorbed onto clay—humus complexes, dissolved in soil water, or taken up directly by plants.

3.2. Water uptake

Chemical weathering processes of rock and soil particles are constrained by
the degree of access which water has to these particles. Thus, for example,
two blocks of limestone may be equally soluble but if one is fissured and
jointed, this block will permit greater access of water and, since chemical
weathering processes largely take place in water, the block will be weathered
to a greater extent than the unfissured, solid block.

Mechanical (or physical) weathering is not considered directly here (see
Trudgill, 1983, Ch. 2 for a more detailed discussion) but its significance lies
in the way in which it breaks up larger rock masses to yield many smaller
particles with large surface areas upon which chemical processes may act.
This can be particularly important where soil formation is taking place
directly over bedrock. Here, the primary structure of the rock, such as the
presence of joints and bedding planes, is important both in allowing access to
mechanical agencies (such as frost and plant roots) and also access of water
for chemical weathering. Of equal importance in these processes is the
character of the rock between the joints and bedding planes in terms of its
porosity (total pore space) and permeability (ability of the rock to transmit
water through the connected pores).

Where the soil parent material is a sedimentary deposit which has been
previously broken down into smaller particles (such as a glacial till or an allu-
vial deposit), then mechanical weathering processes become less important
but the smaller-scale porosity and permeability of the particles become
significant factors in determining the access of water.

Both bedrock weathering and soil particle weathering are important in the
release of nutrients into soils, the former at the base of a soil profile where
weathering bedrock exists and the latter throughout mineral soil profiles. The
two different scales of water uptake—bedrock and particle—are illustrated
in Fig. 3.2. Crystal structure is important at the smaller scale, with the
penetration of water—and chemical weathering agencies—along crystal
surfaces.

Water uptake has been studied by Goudie *et al.* (1970) who measured the
weight increase over time of rock tablets immersed in water. His data for a
variety of rock types are illustrated in Fig. 3.3. Clearly, the most permeable
rocks take up the most water, especially chalk and the more porous sand-
stones. Such variations in water uptake are important in influencing variations in chemical weathering and the simple chemical considerations of
susceptibility to weathering agencies, such as mineral solubility, are, in themselves, insufficient to predict how rocks and minerals will actually weather in
nature. In particular, the permeability of particles and rocks has a marked
influence on the removal of chemical weathering products and therefore on
the continued progression of chemical weathering processes. This topic will
be returned to in terms of the differential movements of ions through the

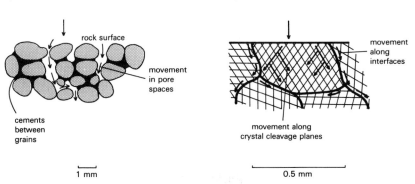

FIG. 3.2. Uptake of water is an important precursor of chemical weathering, as illustrated for water movement: (a) between bedrock particles; and (b) within particles at the crystal scale

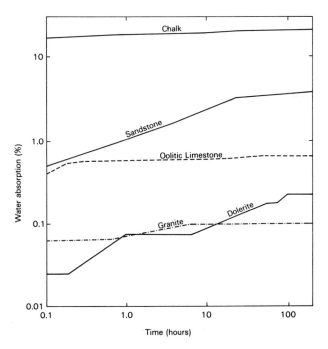

FIG. 3.3. Water uptake over time for 5 rocks of differing porosity, note logarithmic scale, with rapid, higher uptake for porous chalk and low and slow uptake for tightly-packed granite and dolerite rocks (adapted from Goudie *et al*. 1970)

solid matrix (Section 3.3.2), the transport of ions in a moving liquid (Section 3.3.3) and in terms of the leaching of chemical elements in soils (Chapter 7). The fundamental point here is that chemical weathering largely proceeds through the medium of water and therefore the degree of contact between the mineral phase and the liquid phase is obviously an important consideration.

3.3. The basic processes of chemical weathering

Minerals in contact with liquids may be subjected to four sets of processes involving the movements of chemical elements:

1. Reactions at the solid/liquid interface and the movement of ions through the liquid.
2. Differential movements of ions through the solid matrix to the solid surface.
3. Transport of ions in moving liquid.
4. Abstraction of chemical elements by plant roots.

These will be discussed in turn.

3.3.1. Interface reactions and the movement of ions through liquid media

There are five distinct processes which may occur at the solid/liquid interface (Mercado and Billings, 1975):

1. The transport of reactants from the liquid to the interface between the mineral and the liquid.
2. The adsorption of the reactants on the surface of the solid.
3. Chemical reaction on the surface.
4. The desorption of the products of reaction from the surface.
5. The transport of the products from the mineral/liquid interface into the liquid.

These processes are illustrated in Fig. 3.4. Any one of these processes may determine the over-all rate of chemical weathering. Reactions may be controlled by processes 1 and 5 and in that case they are termed transport controlled reactions. Alternatively, they can be controlled by processes 2, 3, and 5 and are termed chemically controlled reactions. However, the over-all reaction rate can be controlled by both sets of processes operating simultaneously.

Transport control. Basically, the controls are the rates at which the reactants and products diffuse to and from the mineral liquid interface. The diffusion of the products from the interface is related to the concentration gradient of the ions in question, the transport distance, the ease with which

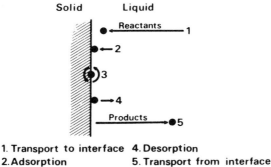

1. Transport to interface   4. Desorption
2. Adsorption   5. Transport from interface
3. Reaction

FIG. 3.4.   Basic reaction processes at the solid/liquid interface

the molecules may move through the liquid, and the surface area of the solid/liquid contact. This can be conveniently expressed in a shorthand form in an equation developed by Nernst (1904):

$$\frac{dc}{dt} = k_d(C_s - C) \tag{3.1}$$

where C_s = concentration of the product in solution at equilibrium
        C = concentration of the product at a given time, t

and thus

$\dfrac{dc}{dt}$ = the change in the concentration of the products in solution over time.

k_d = the transport rate constant. This is derived from a secondary equation:

$$k_d = \frac{D}{\delta} \cdot \frac{A_s}{V}$$

where D = the molecular diffusion coefficient
        δ = the transport distance
        V = the volume of the solution which is in contact with the surface area, A_s, of the dissolving mineral.

The greater the difference between C_s and C, then the steeper the concentration gradient will be and the faster the movement from the solid to the liquid will be (assuming that k_d remains constant).

Active mass.   As well as transport, the active masses of the reactants and products are important considerations. The active mass is that actually taking part in the reactions. The significance of the mass of the products and reactants is expressed in the *Law of Mass Action* (see Gardiner, 1969). The

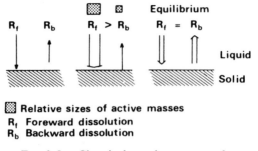

⊞ Relative sizes of active masses
R_f Foreward dissolution
R_b Backward dissolution

FIG. 3.5.   Chemical reactions on a surface

law is that *the rate of chemical reaction is proportional to the active mass of the participants*.

Dissolution processes consist of movements in two directions. There is a forward dissolution, involving the reactants, and a backward dissolution, involving the products. The forward and backward dissolution rates (R_f and R_b respectively) depend upon the active masses of the reactants and products respectively. At equilibrium $R_f = R_b$. The chemical reactions on the surface are illustrated in Fig. 3.5.

Weathering reactions.   From the above two descriptions of the Nernst equation and of the Law of Mass Action it can be seen that weathering reactions will be controlled primarily by the supply of the reactants, the rate of removal of the products, and the surface area of the mineral/liquid contact.

Also involved will be the actual physico-chemical nature of the reactive substances and the mineral in question. These will be discussed further below. The above account of dissolution processes is a simplified one and fuller accounts can be found in Coetzee and Ritchie (1969), Garrels and Christ (1965), while the fundamental reactions and controls upon the chemistry of rock weathering are usefully summarized by Curtis (1976b). Chemical equilibria in natural waters are described by Stumm and Morgan (1970) and the physico-chemical behaviour of nutrients in the soil by Greenland and Hayes (1981), Lindsay (1979), and Bohn *et al.* (1979).

3.3.2. *Differential movements of ions through the solid matrix*

The weathering of some minerals possessing a complex chemical composition does not necessarily proceed by the simple loss of chemical elements from the surface. It is common for the more readily dissolving components to be lost preferentially from within the mineral, leaving behind a matrix of the more slowly soluble components. For example, with a potassium felspar, $KALSi_3O_8$, the potassium may be dissolved out preferentially, leaving behind an alumino-silicate layer at the surface of the mineral (see Mohr *et al.*, 1972,

p. 420). At a second stage the silicon may be removed, leaving an aluminium oxide residue.

Initially, the rates of dissolution and preferential removal of the more soluble components are quite high in fresh rocks but, as the outer layer of depletion increases in thickness, the rate slows down. The rate of diffusion through the less soluble matrix becomes the controlling factor. This, then, can be regarded as a special case of a transport controlled solution reaction.

The depleted layer may itself be dissolved eventually or it may remain in the soil as a residual weathering product. An example of this is given by Berner (1971, pp. 172–3). From an initial felspar, albite ($2NaAlSi_3O_8$), it may be possible to produce kaolinite ($Al_2Si_2O_5(OH)_4$) as a weathering product in the soil, the more readily soluble component, sodium, having gone into solution in a weathering reaction involving water.

The significance of these types of reactions in the study of inputs to the available nutrient store is that it is not necessarily possible to establish which elements will be yielded into solution from a simple study of mineral composition. Some elements may be released preferentially.

3.3.3. *The transport of ions in a moving liquid*

It is clear that, to a large degree, the progression of weathering reactions depends upon the rate at which the reactants and the products can diffuse through a liquid medium. If, however, the liquid medium itself is moving it can readily be seen that the reactions may be considerably facilitated. Mobile soil waters can bring fresh supplies of reactants to reaction sites and they can remove the products of reactions. Referring again to the Nernst equation, it can be seen that in the expression $C_s - C$, C will be kept low by a continual flow of water which is removing weathering products. This will naturally facilitate the transport of ions from the solid phase to the liquid because a steep concentration gradient will be maintained. In contrast, in static waters C will tend to approach C_s until R_f will equal R_b as net dissolution tends to zero.

In soil systems there will be pore spaces of differing sizes and the soil water will tend to be static in the small pores (because of surface tension) and only moving in the larger, interconnecting pores. This does not mean, however, that only active dissolution can take place in the minerals surrounding the larger pores. Chemical reactions can occur in the water in the small pores and the ions can diffuse through the static water (see, for example, Nye, 1972) till they contact the mobile water. This topic is returned to in greater detail in Chapter 7 which is concerned with the leaching of elements in mobile soil waters.

3.3.4. The abstraction of elements by plant roots

This topic is dealt with in greater depth in Chapter 7 and in Section 3.4.2 but it is worth mentioning at this stage that the abstraction of elements from soil water by plant roots can have a marked effect on chemical reaction rates. Simply expressed, the abstraction will have the effect of increasing the concentration gradient by decreasing C in the expression $C_s - C$, thus increasing the dissolution rate.

3.4. Reactants and cation exchange

3.4.1. Hydrogen ions and carbon dioxide

Reactive substances, or reactants, provide the motivation for chemical weathering processes. In some cases the reactant may be oxygen and mineral elements such as iron may combine with it to form iron oxide. In other cases it may be one of the atmospherically or organically derived anions, such as chloride or nitrate, and cations such as calcium may combine with these. The anion reactions are discussed in Section 3.4.4. and in Chapter 4; oxidation and reduction is discussed in Section 3.5.2. While these reactants are often of considerable importance, often the most significant reactant is the hydrogen ion, H^+.

pH.   The concentration of hydrogen ions in soil water is expressed in terms of pH. pH is the negative logarithm to the base 10 of the hydrogen ion concentration:

$$pH = -\log_{10}[H^+] \tag{3.2}$$

where square brackets, [ ], are used to indicate concentration. At neutrality, H^+ ions are balanced by OH^- ions and both have a concentration of 0.0000001 g l^{-1} which is reduced by equation 3.2. above to a simple value of pH 7. Higher levels of acidity are expressed by a lower pH value, for example 0.0001 g l^{-1} becomes pH 4. Since the H^+ is involved in many weathering reactions, as shown below, measurements of the pH of the soil solution are useful indicators of the weathering potential present in soils (see Trudgill, 1983, p. 62.)

Hydrogen ions.   In many weathering reactions the hydrogen ion in solution reacts with a cation in the mineral, replacing it in its position in the crystal lattice of the mineral. The end-result of the reaction is the presence of the cation in solution and the presence of the hydrogen ion in combination with a mineral component.

    The hydrogen–mineral compounds are often very unstable and this contributes to the further breakdown of the mineral, the hydrogen then moving into solution in combination with an acid radical derived from the mineral. The distinction is made by Keller (1957) between (1) cation replacement and (2) combination of mineral anions and cations by hydrogen ions and OH^- ions respectively.

An example of cation replacement and mineral instability is given by Berner (1971, pp. 170–1). Aqueous hydrogen ions combine with potassium felspar, yielding a hydrogen–felspar compound on the surface of the mineral and potassium ions in solution. The hydrogen felspar is unstable and breaks down to yield silicic acid, H_4SiO_4, into solution, leaving an aluminium-rich surface layer behind (Fig. 3.6).

The primary source of hydrogen ions is from the dissociation of acids (the definition of an acid being simply that it is a substance which dissociates in water to yield hydrogen ions) as in hydrochloric acid where HCl yields H^+ and Cl^-, Cl^- being the acid radical.

There are two important sources of acids in soils. One is the dissociation of carbon dioxide in water (Kern, 1960) and the other is from organic acids (see Section 3.4.3). For carbon dioxide the reaction is:

$$CO_2 + H_2O \rightarrow H_2CO_3 \text{ (carbonic acid)}.$$

Carbonic acid is unstable in water and dissociates:

$$H_2CO_3 \rightarrow H^+ + HCO_3^-$$

In the case of organic acids the acid radical is often complicated but the reaction can conveniently be written in a generalized way thus:

$$H(X) \rightarrow H^+ + (X)^-$$

Furthermore, organic acids can react with bases to yield carbon dioxide and water (see Section 3.4.3).

Water itself is partly dissociated and both H^+ and OH^- ions are present as

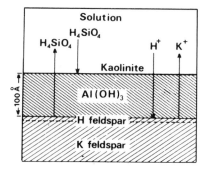

FIG. 3.6.   Diagrammatic representation of felspar weathering (from Berner, 1971)

Source: 'The Genesis and Morphology of the Alumina-Rich Laterite Clays', G. D. Sherman, pp. 154–61, *Problems in Clay and Laterite Genesis*, © AIME, 1952

well as the hydroxonium ion, H_3O^+ (Gold, 1956). The hydroxonium ion can yield hydrogen ions when reacting with minerals.

Minerals can react with hydrogen ions which are present in water, even if the water is not acidified by carbon dioxide. The term hydrolysis is sometimes restricted to reactions with hydrogen ions in pure water and sometimes it is used to indicate the reaction between hydrogen ions and minerals, irrespective of source. Berner (1971, p. 170) restricts hydrolysis to cases where vegetation is absent but Ollier (1975, p. 34) includes discussions of living plants under the heading of hydrolysis. As it is difficult to distinguish sources of hydrogen ions in the soil it is probably most useful to group all the reactions under the term hydrolysis. The qualification should be made, however, that this refers to hydrogen ions dissociated in water and these can occur in pure water without the presence of carbon dioxide or organic acids.

Hydrogen ions can be absorbed into clays and clay-humus complexes just as any other cation can. If hydrogen ions dominate the exchange complex then the term acid clay is used. Hydrogen ions are also present on the surface of plant rootlets. In this case there is active involvement between plant nutrition and weathering (see Sections 3.3.4 and 3.4.2). Cation exchange takes place at the root surface, the root yielding hydrogen ions into solution, and to the clay adsorption sites, while mineral cations move in exchange from the clays and solution into the root.

Carbon dioxide.   Within the soil the sources of carbon dioxide are many but they are all basically respiratory in origin (De Jong and Schappert, 1972; Russell, 1973). The respiring organisms involved are the soil fauna, the plant root biomass, and the bacteria (Gray and Wallace, 1957) and fungi which are actively engaged in the decomposition of organic matter. While carbon dioxide is present in the atmosphere above the soil, the amount present (0.003 per cent) is very small when compared with the amounts which can occur in soil air. The concentration is commonly in the range of 0.5–5 per cent but it can be as high as 10–20 per cent. Measurements of carbon dioxide levels in the soil air are commonly taken as indications of weathering potential on the assumption that the carbon dioxide in soil air is in equilibrium with that dissolved and dissociated in soil water (Eriksson, 1952; Sweeting, 1972, pp. 31–4).

The identification of separate sources of carbon dioxide in the soil is difficult but Garrett and Cox (1973) estimated that most of the carbon dioxide evolved from a forest floor in Missouri was contributed by root respiration and associated micro-organisms. Both this study and many other studies (for example, Kononova, 1966; Tamm and Krzysch, 1963; Reiners, 1968; Edwards and Sollins, 1973; Boynton and Compton, 1944) stress that carbon dioxide production is dependent upon temperature and moisture. This is because these factors control biological productivity. This is of interest since it suggests that weathering potential can be predicted from a study of environmental factors which influence biological productivity. This point is examined further in Sections 3.4.2 below, and 3.4.3.

3.4.2. Root uptake and cation exchange

Rhizosphere processes are important influences on mineral weathering input because they provide reactants and they remove products. Moreover, the processes will be directly related to plant productivity and therefore they can be modelled in terms of the type of plant involved and of the environmental factors which are influencing their growth. It is suggested by Steenbjerg (1954) that, all other things being equal (i.e. in controlled pot experiments), plant growth and nutrient release are directly related.

Root processes. Roots are instrumental agents in the weathering of minerals not only because they take up cations in exchange for hydrogen ions (Wiklander, 1964; see also Sections 3.3.4 and 3.4.1) but also because they actively produce exudates (chiefly sugars and proteins) which act as substrates for the nutrition of soil micro-organisms. The rhizosphere micro-organisms act as agents of mineral solubilization in the rhizosphere (Richards, 1974). The micro-organisms appear to attack the minerals by the production of organic acids, some of which may act to chelate mineral cations (see Section 3.4.3).

Biologically motivated weathering systems are also described by Keller and Frederickson (1952). They suggest that clay colloids occupy an intermediate position between the roots and the minerals (Fig. 3.7(a)). Hydrogen ions move from the rootlets towards the mineral down the concentration gradient, occupying adsorption sites on the clays *en route*. Hydrolytic weathering occurs at the mineral/clay contact and nutrient cations move in the opposite direction, being actively taken up by the plant. Richards (1974) proposes a mechanism which does not necessarily involve the intervention of a clay colloid but simply involves the output of CO_2 and cation exchange (Fig. 3.7(b)).

Two-way soil-plant relationships. If the processes described above are in operation it can be expected that the presence of plants will considerably modify soil characteristics. This is of interest since it is often tempting to view soil characteristics as a simple causative factor in influencing plant distributions (for a detailed discussion see Rorison, 1969).

One of the clearest bodies of evidence for the modification of soil by plants is from the work of Grubb *et al.* (1969) and Grubb and Suter (1971) on the effect of heather, *Calluna vulgaris*, on soil acidity. Transects under *Calluna* plants on a chalk heath in southern England clearly show a marked acidity under the plant (Fig. 3.8). Whether this is due to root action or to the acidity of litter is not entirely clear but the action of roots and root micro-organisms

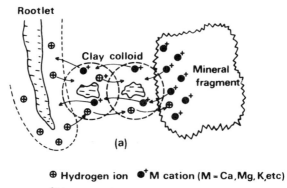

⊕ **Hydrogen ion** ●⁺**M cation (M = Ca,Mg, K,etc)**

◌ **Oscillation volume of colloid and cations**

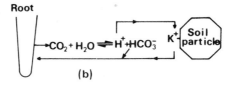

FIG. 3.7. (a) Diagrammatic representation of cation movement between rootlets, clay colloids, and mineral fragments (from Keller and Frederickson, 1952), (b) Diagrammatic representation of the carbon dioxide hypothesis (from Richards, 1974)

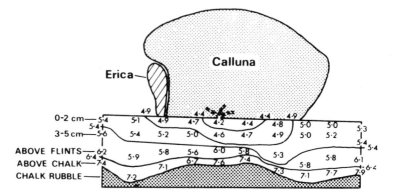

FIG. 3.8. pH transect under a plant of *Calluna vulgaris* with a suppressed plant of *Erica cinerea*. Horizontal scale same throughout and same as vertical scale above soil. Vertical scale within soil exaggerated to 2.5 times the horizontal. Height of bush was 55 cm and pH determinations were 15 cm apart (from Grubb *et al.*, 1969)

have been demonstrated by Duff *et al.* (1963) and by Boyle and Voigt (1973). In the latter work the root systems of pine seedlings were found to be able to release potassium from biotite. The former reports a similar release of potassium from micas and felspars by bacteria which produce 2-ketogluconic acid in the root regions of crop plants. Microbial solubilization of a granite sand has been demonstrated by Berthelin and Dommergues (1972).

Fungal hyphae such as those of *Aspergillus niger* (Eno and Reuszer, 1950) can be important in solubilizing minerals (Henderson and Duff, 1963). Fungal hyphae can be found in association with the roots of heath plants (Gimingham, 1972) and with conifers (Handley, 1954). These associations are termed mycorrhiza and it is possible that they afford some facility for mineral solubilization. There is considerable implication that soil acidity may be enhanced in the rhizosphere by their action.

Plants may possibly influence soil chemistry by the possession of chelating mechanisms which aid the preclusion of uptake of very soluble elements present in toxic concentrations (Grime and Hodgson, 1969, pp. 94–6; Jeffries *et al.*, 1969). Moreover, where some elements are in a freely soluble form, such as iron and manganese in waterlogged, reduced conditions, plants may exhibit a reduced transpiration rate in order to prevent the excessive uptake of elements present in toxic amounts (Jones, 1971). While these mechanisms appear to exist, they seem to be an adaptation to soil type and their effects upon soil chemistry are liable to be small.

Modelling. A model for mineral solubilization in the rhizosphere is shown in Fig. 3.9. Notice that the exchangeable and soluble ions can be produced both by simple inorganic solution process and also by the actions of rhizosphere micro-organisms.

One of the difficulties in detailed modelling of the root uptake system is

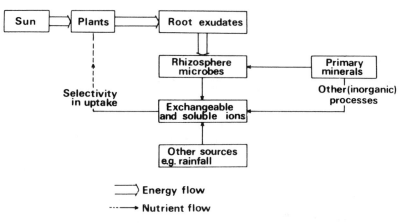

FIG. 3.9. A model of nutrient pathways in the rhizosphere (after Richards, 1974)

that plants vary in their capacity for cation uptake. It is suggested by Richards (1974, p. 181) that dicotyledonous roots have, as a general rule, a higher cation exchange capacity than monocotyledons. Furthermore, plant uptake is not a simple matter of the uptake of all the available elements. Some selectivity is involved and, for example, Russell (1973, pp. 550–1) suggests that plants often appear to contain more phosphorous and potassium in proportion to calcium than the soil does. Thus a knowledge of individual plants and their behaviour with respect to each element is necessary in order to construct detailed models. While this is attempted for silicon and calcium in Chapter 7 at this stage in the discussion only a first-order outline of plant-facilitated weathering can be attempted.

Simply, the basic assumptions are made that plant-facilitated weathering is a function of root uptake and also that root uptake is a function of plant productivity. It would thus be expected that plant productivity data would be a guide to the extent of plant-facilitated weathering. Data given by Odum (1962) suggest that a tropical forest has a productivity of 10–25 g m^{-2} per day while in temperature forest or grassland the magnitude is in the order of 3–10. Clearly, this suggests that the optimum for biological weathering would be in a hot, moist climate where it could be expected that it would be about 2 to 8 times higher than in a temperate climate. The evidence for the rigorous proof of this kind of pattern is scarcely available, however. The greatest challenge in modelling lies in the difficulty of establishing the relative importance of all the different processes which have been discussed so far. For instance, Mitchell (1973) demonstrated that bracken (*Pteridium*) rhizomes can release phosphorus from inorganic sources, but whether the agent is root exudates or a decomposition product is not known. If all the reactants for weathering were derived from one source, the factors governing this source could be modelled and then the degree of biologically induced weathering could be predicted for any one given set of conditions. However, the sources for reactants are many. In summary, they are root respiration, soil fauna respiration, organic matter decomposition, root uptake, rhizosphere solubilization by micro-organisms, and direct root exchange. However, while these are all difficult to model separately, they are all basically biological processes and thus can usefully be modelled in broad terms from a knowledge of the environmental factors which govern biological activity. Thus the processes can be expected to be at a maximum under warm, moist conditions but at a minimum if either moisture content, or temperature, or both, are too low.

The element of feedback in rhizosphere process systems is probably considerable. It is suggested by Richards (1974, p. 180) that plant productivity encourages the provision of root exudates which will encourage microbial solubilization of minerals. In turn, the release of elements will help to increase plant productivity. In terms of root activity, the availability of nutrients in the soil store and the release of nutrients from minerals are thus intimately related.

3.4.3. *Organic acids, organic matter decomposition, and chelation*

While the motivation for mineral solubilization by plant roots can be clearly seen—that of direct, active uptake—there are other very important organically influenced weathering processes in the soil which are rather less active in terms of higher plant functions. These processes are related to the decomposition of plant and animal remains in the soil by the soil bacteria, fungi, and fauna. The decomposition processes release various substances which have a capacity to facilitate the weathering of soil minerals. The role of organic matter decomposition as a factor in rock weathering is described by Manskaya and Drozdova (1969) and Ong *et al.* (1970). The role of mobile soil water is vital in this context as it carries the organic substances from the upper layers of the soil, where they are actively formed, to the lower, mineral layers of the soil.

Basically, the organic substances have three functions: the production of hydrogen ions by direct dissociation, the production of carbon dioxide, and chelation (Fig. 3.10). The former two have been mentioned briefly above. Chelation is the combination of an organic compound with a metal cation, the metal actually entering into the chemical structure of the organic compound.

Chemically speaking, a chelate is a type of complex. A complex is defined as a particle that is formed by the association of two or more simpler particles and which can exist in solution. The composing particles can be molecules or ions or other complexes. Complexes can be chelates or ligands. Ligands can be $1-$, $2-$, or more polydendate, i.e. it can have 1, 2, or more atoms acting as co-ordinating atoms in the chemical structure. If a polydentate ligand is bound to one central atom by two or more 'dents' ('teeth' or linkages) then the complex is called a chelate (Mohr, van Baren, and Schuylenborgh, 1972).

The chemistry of the decomposition of organic matter is extremely complicated and difficult to study. There is every gradation from complete

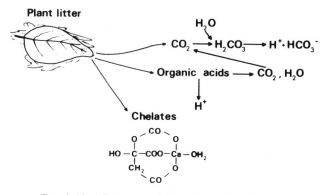

FIG. 3.10.   Decay products of organic substances

biological structures like leaves, through half-decomposed organic matter and various organic acids, down to simple carbon dioxide, water, and nitrogen. Substances which can be recognized as organic compounds are often large and complicated in molecular terms. For example, Schnitzer and Desjardins (1962) were able to prepare an organic matter fraction from the Bh horizon of a podsol soil, though there was considerable difficulty in obtaining a pure enough extract. The compounds appeared to have a mean molecular weight of 670.

A good review of humic substances in the environment is provided by Schnitzer and Khan (1972). Other useful references are those of Beckwith (1955), Breger (1963), Calvin and Martel (1953), Chaberek and Martel (1959), Goring and Hamaker (1972), Grobler and Pauli (1964), Handley (1954), Manskaya and Drozdova (1968, 1969), Martel and Calvin (1952), Murray and Love (1929), Schnitzer and Skinner (1963), Singer (1973), and Swain (1970). The humic substances found in the soil are many and include humin, fulvic acid, humic acid, and hymatomelanic acid. These are not necessarily closely defined substances and are often only identifiable by fractionation techniques (Fig. 3.11). Other organic derivatives include polyphenols, hydroxy and phenolic acids, carboxyls and phenolic hydroxyls. The simpler substances include a whole range of acids, for example oxalic, citric, malic, succinic, gallic, aspartic, salicylic, and tartaric acid as well as mono-, di-, and trihydroxy-benzoic acids. Some of these are derived directly from recognizable higher plant substances, for instance gallic acid is derived from tannin, but in other cases the derivation is not necessarily clear.

Many of the above substances participate in organo-metallic reactions.

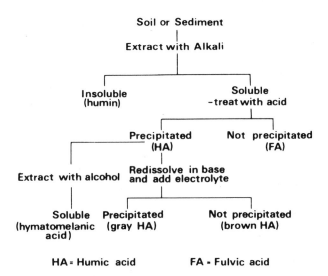

FIG. 3.11.   Fractionation of humic substances (from Schnitzer and Kahn, 1972)

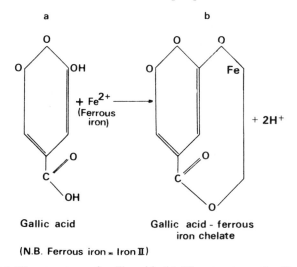

FIG. 3.12. (a) The structure of gallic acid, (b) The structure of gallic acid–ferrous iron chelate (from Mohr *et al.*, 1972)

These are described in detail in Schnitzer and Khan (1972, pp. 203–51) and also by Bremer *et al.* (1946). Metal-complex formation by lichen compounds is described by Iskander and Syers (1972). The topic is also dealt with in a soils context by Davies (1971) and in Mohr *et al.* (1972, pp. 451–9). Fulvic acid-metal ion reactions are described by Gamble and Schnitzer (1973) and Hawarth (1968) suggests that soil humic acid is composed of an aromatic 'core' to which metals are attached by a chemical and/or physical means. The chelation of many elements derived from mineral weathering is attested to in these texts, especially the chelation of iron. For example, citric acid forms a 1:1 complex with ferric iron and p-hydroxybenzoic acid also forms similar complexes. Gallic acid forms complexes with ferrous iron and the nature of the complexing action is illustrated in Fig. 3.12. As an example of chelation of other cations, citric acid is also known to form a calcium citrate (hydrate) chelate, as illustrated in Fig. 3.13. Note that in all these illustrations the iron

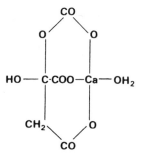

FIG. 3.13. The structure of a calcium citrate (hydrate) chelate (from Swain, 1970)

and the calcium are intimately involved within the chemical structure of the organic compound and thus the process is substantially different from the dissolution process described above (though in part the action of chelates also involves dissolution because the compounds are acids).

It is clear that any chelates which are present in mobile soil waters will play an important role in the transport of mineral cations down the soil profile. The chelates tend to be formed in the upper organic and more acidic soil horizons and as they are washed down through the lower mineral horizons they pick up cations. They tend to become increasing insoluble as more mineral elements are complexed. Thus they tend to become deposited lower down the horizon as they become less soluble and also partly hydrolysed. The coagulation of humic colloids by metal ions is described by Ong and Bisque (1968).

In the case of iron there is evidence (Bloomfield, 1963) that simultaneous reduction and chelation occurs in the upper soil horizons, mobilizing iron and facilitating its elution to lower horizons. The organic acids produced in the upper horizons are soluble and mobile. They can pick up considerable amounts of metals, especially iron and aluminium. As the acids are moved down the soil profile under the influence of rainfall they tend to become increasingly saturated with iron and aluminium, become less soluble and precipitate. It is in this way that Bs, Bfe (Bi), and Bh horizons form towards the base of the soil profile.

These processes have been simulated by Atkinson and Wright (1967) using an artificial chelate (EDTA) in soil column experiments (Fig. 3.14) and by

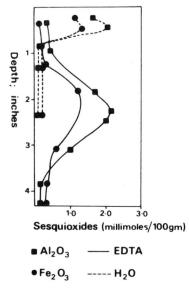

Fig. 3.14.   The distribution of sesquioxides extractable from a soil profile by dilute acid after leaching with water and a chelate, EDTA (from Atkinson and Wright, 1967

Kerpen and Scharpenseel (1967) using isotopically labelled humus. Several experiments and observations have been undertaken by Bloomfield (1950, 1951, 1952, 1953a, 1953b, 1954a, 1954b, 1954c, and 1963). His experiments demonstrated the likelihood of the mobilization of iron and aluminium by organic derivatives from leaves and bark of a number of tree species. Similarly, Schnitzer (1959) has extracted leachates from decomposing leaves washed with rainfall. The leaf leachates from maple, poplar, birch, and beech were effective in taking up iron in that descending order. The work of Coulson *et al.* (1960a and b) related soil base status to the nature of poly-phenols in litter. The leaf polyphenols were able to chelate ferrous iron. The litter from beech trees growing on base deficient sites was richer in simple polyphenols than those growing on base rich soils.

The amounts of cations solubilized by chelate action are usually consid-erably greater than those solubilized by the actions of water and hydrolysis alone. Moreover, the ratios in which cations are solubilized to form organic complexes often show marked contrasts with the ratios of cations solubilized in water. In addition, the ratio of the solubilization of individual elements may change in each organic substance. These types of phenomena have been demonstrated by Huang and Kiang (1972) and Huang and Keller (1972). They shook a selection of felspars in 0.025M solutions of organic acids and compared the dissolution process which occurred in several acids with those which occurred in water. The organic acids dissolved far more calcium, sodium, silicon, and aluminium from plagioclase felspars than water did. Equilibrium concentrations ranged from 1 to 100 μ mol l^{-1} in water but from 50 to 1000 μ mol l^{-1} in organic acids (Fig. 3.15). The action of strongly complexing organic acids, such as salicylic acid, was compared with that of

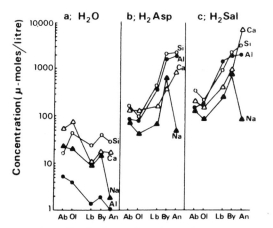

FIG. 3.15. Comparative dissolution of cations from a series of minerals in (A) deionized water, (B) aspartic acid, (C) salicylic acid. Values for experimental period of 24 days at room temperature. Ab: albite, Ol: oligoclase, Lb: labradorite, By: bytownite, An: anorthite (from Huang and Kiang, 1972)

weakly complexing ones such as aspartic acid. The results are summarized in Table 3.1. The dominant element of two pairs, Ca and Na or Si and Al, is shown. Notice that calcium dominates over sodium in both water and organic acids. However, in water silicon is dominant over aluminium but the picture is reversed, or aluminium is at least equal to silicon, in the organic acids. Aluminium is thus solubilized more in organic acids than in water. Organic acids can thus substantially affect the amounts and ratios of elements going into solution from minerals.

While chelation processes were mentioned in the context of root processes (Section 3.4.2) there can be a great difference in the effects of root chelation and the organic matter decomposition processes described above. The effects of the root processes are more or less confined to the rhizosphere while the effects of chelation by the products of organic matter decomposition may be seen throughout the whole soil profile.

In net effect, the two processes can, in fact, be viewed as acting somewhat in opposite directions. Leaching by the action of chelates (cheluviation) depletes the upper soil horizons of metal cations while uptake by plant roots retains nutrients and preserves them within the soil–vegetation system. It can thus be emphasized how important plant uptake is in retaining nutrients within the over-all system against the forces of nutrient loss. In addition, with regard to chelation in immature soils, rhizosphere activity can take place as soon as plants have begun colonizing soils whereas it is necessary for organic matter to begin to accumulate and decompose, at least on the surface of the soil, for cheluviation to occur.

Modelling.   Chelates released by the decay of organic matter act as reactive substances involved in the breakdown of minerals. While the chemical processes involved are rather more complicated than those described in Section 3.3.1, nevertheless the general principles still apply. The over-all control of the system will be by (1) the provision of 'reactants' (chelates), (2) the transport of weathered products, and (3) the rate of reactions between organic and mineral substances (which are, in fact, only poorly understood).

As the breakdown of organic matter is a biological process, it is possible to model the system in terms of biological controls such as temperature and moisture. Moreover, other environmental factors will be important, such as pH. At neutral or alkaline pH values, bacteria will tend to dominate the decomposition biota and, also, earthworms will be abundant. At acid pH values, fungi tend to dominate the decomposition biota. Fungi are recognized for their production of organic acids, such as oxalic and citric acid (Henderson and Duff, 1963). Thus the type of weathering achieved can be altered in essence by the nature of the biota involved (and the substances they produce) and in degree by temperature and moisture. A model of organic matter decomposition and carbon dioxide production is given in Fig. 3.16.

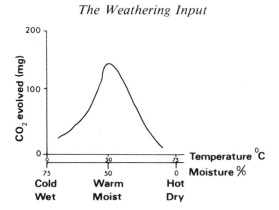

FIG. 3.16. Organic matter decomposition and carbon dioxide production as a function of temperature and moisture (after Kononova, 1966)

3.4.4. Anions in solution

Mineral cations can be combined with anions in solution, as well as being adsorbed onto clay–humus complexes, involved in chelates, or taken up by plant roots. Anions such as NO_3^-, HCO_3^-, SO_4^{2-}, and Cl^- may be present in the soil solution. In addition, NO_3^- and HCO_3^- may be derived from organic processes, such as the decomposition of organic matter, while all four anions cited above can be derived from the atmosphere (see Chapter 4).

The presence of anions in water can considerably alter the physico-chemical properties of water. The function of the anions in a weathering context is that they can combine with the products of dissolution of mineral solids. It is suggested by Jurinak and Griffin (1972) that the nitrate ion, NO_3^-, can react with solid phase calcium carbonate in calcareous soils. It has been demonstrated by McColl (1969) that cation transport through soils can be largely regulated by the production of mobile anions, especially by bicarbonate, formed by the process of organic decomposition. Thus the anions can interact with minerals and they may facilitate the removal of reaction products and the progression of dissolution processes. Mobile soil waters charged with anions will thus create a favourable environment for weathering and leaching processes.

3.5. Mineral stability

It is difficult to specify the stability of a mineral without the specification of the weathering environment in which it is present. Clearly, for example, minerals will dissolve in different ways in the presence or absence of organic acids (see Section 3.4.3). However, it is possible to describe some concepts concerning mineral stability in a general way (Section 3.5.1, below). Besides

Table 3.1 Changes in solubilization ratios in water and organic solvents

Elements are given in pairs of Ca and Na or Si and Al. Note that the dominant element of a pair often changes with solubilization in the organic acids. $>$ = greater than; $\doteq$ approximately equal to.

Mineral	Dominant element of pair in mineral	Dominant element of pair in solvents at equilibrium		
		Water	Aspartic acid	Salicylic acid
1. Albite	Na > Ca, Si > Al	Ca > Na, Si > Al	Ca > Na, Si > Al	Ca > Na, Si > Al
2. Oligoclase	Na > Ca, Si > Al	Ca > Na, Si > Al	Ca > Al, Si $\doteq$ Al	Ca > Na, Si $\doteq$ Al
3. Labradorite	Ca > Na, Si > Al	Ca $\doteq$ Na, Si > Al	Ca > Na, Si $\doteq$ Al	Ca > Na, Al > Si
4. Bytownite	Ca > Na, Si > Al	Ca $\doteq$ Na, Si > Al	Na > Ca, Si $\doteq$ Al	Ca $\doteq$ Na, Si $\doteq$ Al
5. Anorthite	Ca > Na, Si > Al	Ca > Na, Si > Al	Ca > Na, Si $\doteq$ Al	Ca > Na, Si > Al

Note: Percentage composition of minerals

	Na_2	CaO	SiO_2	Al_2O_3
1.	11	0.2	66	22
2.	8	4	59	27
3.	5	9	52	33
4.	4	14	46	34
5.	1	17	42	38

Source: Huang and Kiang (1972)

the weathering factors already discussed in Sections 3.3 and 3.4 one important factor in mineral stability is the oxidation or reduction status (Eh) of the
weathering environment, and this is discussed in Section 3.5.2 below. The
stability of a particular mineral in a given environment not only controls the
release of chemical elements but it also influences the production of weathering products such as clays (Berner, 1971; Ismail, 1970). Clay type is an important factor in influencing cation exchange capacity (as is illustrated in a
simple fashion in Courtney and Trudgill, 1976, p. 30).

3.5.1. Mineral stability

A useful general reference text is to be found in Ollier (1975) but a more up-
to-date exposition is in Curtis (1976a and b). There are two sets of factors
which influence inherent mineral stability. One is concerned with the nature
of the chemical and of the crystal structure of the mineral; the other with the
amount of energy change involved in weathering reactions. Simply stated,
weak bonding and incoherent crystal structures are often characteristic of
minerals prone to decomposition. In terms of energy change, the standard
free energy change of reaction (abbreviated to G_r^0) is the sum of the free
energy of formation of all the reaction products minus the sum of free energy
of the reactants (Curtis, 1976a, pp. 65–6):

$$\Delta G_r^0 = \Delta G_r^0 \text{(products)} - \Delta G_f^0 \text{(reactants)}$$

Assuming that reactions occur in natural environments where the reactants
(such as H^+, O_2, and H_2O) are freely available, ΔG_r^0 will be negative as the free
energy of the reactants will be greater than that of the products (this being a
necessary prerequisite for spontaneous reaction). The greater the negative
value of ΔG_r^0 the less stable the mineral in question will be in terms of the
reaction involved. Table 3.2 demonstrates the relative stability of selected
minerals. In the examples given, muscovite is the most and pyrite the least
stable. During hydrolysis, muscovite will be more stable than the felspars and
during oxidation fayalite will be more stable than pyrite. Further examples
and a fuller account can be found in Curtis (1976a).

Table 3.2 *Standard free energy changes of selected weathering reactions*

Mineral	Reactants	ΔG_r^0, Kcal g atom^{-1}
Muscovite, $KAl_3Si_3O_{10}(OH)_2$	H^+, H_2O	$- 0.32$
Microcline felspar, $KAlSi_3O_8$	H^+, H_2O	$- 0.51$
Anorthite felspar, $CaAl_2Si_2O_8$	H^+, H_2O	$- 1.32$
Fayalite, Fe_2SiO_4	O_2	$- 6.58$
Pyrite, FeS_2	O_2, H_2O	-17.68

Source: Curtis (1976a)

3.5.2. Oxidation and reduction

Mineral stability can depend upon the amount by which the soil is oxy-
genated. This is especially true of minerals which contain iron (Hem and
Cropper, 1959). Oxidized iron is less soluble in soils than reduced iron. The
former is referred to as ferric iron (or iron III as it is trivalent) and the latter as
ferrous (or iron II as it is divalent). Oxidation is, chemically speaking, the loss
of a negative electron and reduction is the gain of a negative electron. Therefore during oxidation a substance becomes less negative, or more positive,
and during reduction it becomes more negative. Thus, oxidation–reduction
status can be measured potentiometrically and this is usually effected by
insertion of a clean platinum electrode into the soil and the measurement of
the small charge induced by the oxidation or reduction of the electrode. These
electrical measurements are expressed on an Eh scale where a positive reading
indicates a degree of oxidation and a negative reading indicates a degree of
reduction. Eh and pH are important factors in biological and chemical
processes (Baas-Becking *et al.*, 1960) and they interact to control the form in
which iron exists in soils (Hem, 1960), as shown in Fig. 3.17. The reduction

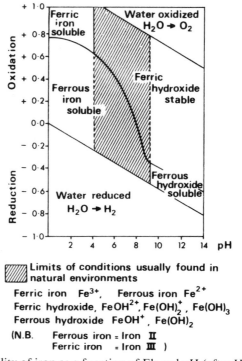

FIG. 3.17.   Solubility of iron as a function of Eh and pH (after Hem and Cropper,
1959)

of iron to the ferrous form renders it soluble and mobile. As has been mentioned above (Section 3.4.3), chelates can play an important role in this, facilitating weathering reactions by simultaneous chelation and reduction. For example, gallic acid (a tannin derivative) may act as an electron donor, ferric iron acting as the electron acceptor:

$$Fe(OH)_3 + e \rightarrow Fe^{2+} + 3OH^-$$
(ferric)   (ferrous)

It is suggested by Mohr *et al.* (1972, p. 456) that iron mobilization can only be achieved when reduction by chelates is involved while Oborn and Hem (1961) stress the importance of microbial factors in the solution and transport of iron. Biological factors are therefore strongly implicated in the question of mineral stability in the context of iron in soils.

3.6. The weathering input: summary and modelling

The weathering input to the soil nutrient store will be influenced by the following:

1. Chemical dissolution processes and factors.
2. Biological provision of chemical reactants and biological uptake of products.
3. Chelation, both in the rhizosphere and by organic derivatives present in mobile soil water.
4. Removal of weathering products in mobile soil waters.
5. The stability of minerals in given weathering environments.

In modelling the over-all input system, one of the interesting considerations is that it can be suggested that mineral breakdown and plant uptake should tend to become mutually adjusted. Indeed, the work of Steenbjerg (1954) suggests that growth of plants and the release of nutrients from mineral sources are directly related. Plants were grown under controlled conditions and the rate of growth correlated closely with nutrient release. The amount of mutual adjustment in the whole system is clearly a topic of great interest and one which will be returned to in Chapter 7. At this stage, however, it will be useful to discuss one further factor, which, although not a weathering input, may have some bearing upon weathering processes. This is that the decomposition of organic matter not only releases chelates and carbon dioxides but also releases the nutrients which were present in the organic tissue. Organic matter returned to the soil surface as leaf litter will contain nutrients which were taken up by the plant roots. For example, leaves will contain chlorophyll which, as well as carbon, hydrogen, oxygen, and nitrogen, also contains magnesium; its chemical formula being $C_{55}H_{72-7}MgN_4O_{5-6}$. Calcium, sodium, and potassium and many other elements are also involved in plant material and they will be released upon the decomposition of organic matter.

The interest in the decomposition of organic matter is in the net effect in terms of released substances, because both substances with the potential for weathering minerals and also nutrient elements are released. It is probable that the cation content of the decomposing organic matter is relatively low since organic matter is usually negatively charged (Gamble and Schnitzer, 1973, p. 265; Black and Christman, 1963a). It is probable that the net effect of decomposition is to release more in the way of weathering potential than of nutrients, which could act to offset the potential. There is some interest in the fate of the released elements. They could become leached out of the soil system, in which case further mineral weathering would be necessary in order to replace them for further plant uptake (unless they were supplied atmospherically). Alternatively, they could be taken up by the plant, in which case compensatory weathering would not necessarily have to occur. The relative importance of these two routes is not clearly established, but in a study of the inputs and exports of nutrients in the Hubbard Brook ecosystem (New Hampshire, USA), Gosz *et al.* (1973) report that the net output from the ecosystem of Ca, Mg, and K in streams is very small when compared to the amounts released by the decay of leaf and branch litter. Apparently, most of the nutrients are retained in the system. This would seem to indicate that leaching losses of released nutrients are minimal and that uptake and recycling are involved. This would tend to minimize the role of primary mineral weathering as a nutrient source. However, there are three qualifying factors. The first is that nutrients could be leached below the rooting zone but not as far as the streams. They would be present in the lower soil horizons but not be recorded as an ecosystem export. Further weathering of minerals in the rooting zone could occur in order to replace those eluviated. Secondly, the retention would probably only be maximized in a mature, undisturbed ecosystem under natural conditions. Under any other conditions, other exports may occur which could be offset by weathering. Thirdly, there is an allied point that as plants grow and colonize a substrate, biomass production will involve the progressively increased uptake of nutrients which would not necessarily be involved in litter return. This will be especially true where nutrients are being stored in perennial woody tissues. In mature ecosystems nutrients will be released from the fall and decay of tree trunks; under developing ecosystems where trees are not yet falling and dying of old age, or where timber is cropped, nutrient uptake is not liable to be balanced by return processes. Thus, under these systems, biologically motivated weathering processes are liable to be very important.

While it is difficult to obtain quantitative evidence for these suppositions above, they can, at least, be usefully itemized in the form of a diagrammatic model (Fig. 3.18). The over-all model is shown together with the various possible routes and flows.

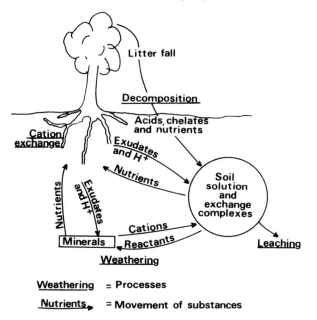

FIG. 3.18. Diagrammatic representation of weathering and related nutrient pathways

Finally, the discussion can be usefully broadened to consider further the question of feedback between the factors so far mentioned. These are the nutrient status and stability of the soil minerals, nutrient availability in the soil nutrient store (adsorbed on clay colloids and in the soil solution), and biological activity. Feedback has already been mentioned in the context of the rhizosphere activity (see Section 3.4.2) but it may also exist in the context of organic matter decomposition. Apart from the factors of moisture and temperature, the degree of decomposition of organic matter will depend upon the influence which litter acidity and nutrient content have upon biological activity (Williams and Gray, 1974). On a nutrient poor soil, the vegetation growing will tend to be comprised of heath or coniferous species. These have a litter which is acid. Figures are given by Handley (1954) for the pH of aqueous extracts of freshly fallen litter as follows: *Calluna vulgaris*, 3.4; *Pinus ponderosa* 3.4; *Pinus contorta* 4.0. Acid litter will tend to be decomposed by fungi because they are more dominant in acid soils (Buckman and Brady, 1969, p. 396). Organic acids will be produced by the fungi and the end-product of this kind of decomposition in acid conditions is a partially decomposed mor humus with a high phenolic content. This will have considerable influence on the type of weathering actions, as discussed in Section 3.4.3, especially in that Al will be preferentially dissolved over Si. The feedback relationship is illustrated in Fig. 3.19.

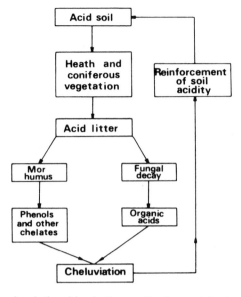

FIG. 3.19.   Feedback relationships in the weathering–nutrient store system

In summary, it is clear that a knowledge of biological activity is crucial to
an understanding of the weathering of minerals in the soil. Although the
trends of the processes involved can be described in chemical terms and by
chemical equations, the motivation is substantially biological. It is true,
however, that water is the medium through which the reactions take place.
Thus it is evident that a study of the moisture characteristics of the soil is also
important and that the influence of the movement of water in the soil is
fundamental to the understanding of the whole system. The study of rainfall
input forms the theme of Chapter 4 and the study of the hydrological
throughput of the soil that of Chapter 5. Clearly, however, mobile soil waters
tend to have an action which is acting partly in opposition to the biological
system of solubilization and uptake in the rhizosphere. Percolating rain-
water acts to bring fresh organic and inorganic reactants to weathering sites
and to remove fresh products. This can occur in the same zone where plant
roots are present. A factor of importance here is the relative amounts of
nutrients held on clay colloids and clay–humus complexes compared with the
amount in the soil solution. Those held in the latter are easily lost in the
mobile soil waters; those held in the former are far less easily lost and are only
readily moved by cation exchange, which is the primary mechanism of
nutrient uptake by plants. The systems involved can be modelled as shown in
Figs. 3.20 and 3.21.

Weathering processes in soils are thus a vital input to the soil nutrient store.
The nature of the input can be considerably influenced by biological activity

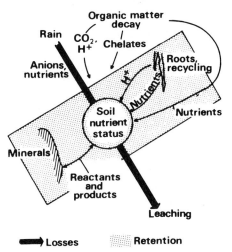

FIG. 3.20.   Summary diagram of the opposing forces—weathering, leaching uptake, and return

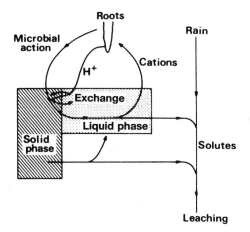

FIG. 3.21.   Summary diagram of reaction, retention, exchange, and losses

and by water movement. The relative importance of the input supply depends upon the importance of the other inputs and outputs discussed in the next three chapters. The topic of the mutual interrelationships of the relevant factors is returned to again in the discussion in Chapter 7.

Further reading

The weathering input; soil plant relationships, soil organic matter; soil chemical processes. (Bear, 1964; Black, 1968; Clarkson, 1974; Deju, 1971; Fried and Broeshart, 1967; Hallsworth and Crawford, 1965; Hatteri, 1973; Kononova, 1966; Loughnan, 1969; Lukashev, 1970; McLaren and Peterson, 1967; McLaren, Peterson, and Skujins, 1971; Nye and Tinker, 1975; Ollier, 1975; Rorison, 1969; Stumm and Morgan, 1970; Sutcliffe and Baker, 1974; Trudgill, 1983.)

4   The Atmospheric Input

Our trouble is that when we make a single hypothesis, we become attached to it.

CHAMBERLAIN

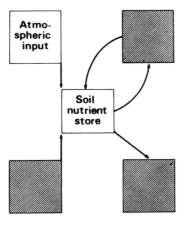

FIG. 4.1.

4.1. Introduction

Rain-water contains chemical elements in solution and, although the quantities involved are usually very small, they can form a very important input to the soil nutrient system (Swank, 1984; Swank and Henderson, 1976). Particulate matter can also arrive at the surface of the earth by dry fallout or it can be washed from the atmosphere as particles in suspension in rain drops. The particles may, in certain circumstances, constitute the nuclei around which precipitation droplets form. The type of precipitation is not considered here; what is of interest is that precipitation can contain both plant nutrients and also ions which may become involved in weathering processes (Carroll, 1962; see also Section 3.4.4).

The atmospheric supply of nutrients is usually small relative to the supply of nutrients found in mineral soils. The atmospheric supply only becomes of major significance in the case of infertile, low nutrient content parent materials (such as quartz sands) and in the case of ombrogenous, raised peat bogs. These receive no drainage waters from surrounding rocks and the soil mineral matter is isolated from the growing plants by considerable thicknesses of peat.

In recent years, the acidity of rainfall and dry deposition have become topics of increased interest. The literature on these topics is immense but useful summaries are presented by Cryer (1986) on the physical and chemical processes involved and by Mason (1986) on the issues involved. The effects on soils are reviewed by Tabatabai (1985) and also in the first issue of the Journal *Soil Use and Management*, (Blackwell Scientific Publications), Vol. 1, No. 1. The effects of acidic atmospheric inputs on soil and vegetation systems are discussed in Section 4.4, especially in terms of soil acidification and the mobilization of nutrients.

4.2. Precipitation chemistry

The amount of each individual ion which is present in precipitation and dry fallout varies from place to place, especially with proximity to the sea (Yaalon and Lomas, 1976), the provenance of rain-bearing winds, and the pattern of daily weather (Gorham, 1958). In stormy weather considerable amounts of sea spray can be dispersed into the air and it can be blown several kilometres inland. Winds with continental provenances frequently have a higher dust content than those with maritime provenances. Local pollution sources can have considerable effects, especially in the case of quarry dust or road dust and also of industrial emission, where sulphuric acid and carbon, amongst other substances, are often involved.

Accordingly, it is difficult to generalize about the composition of precipitation, but several published accounts do exist which can be used to give an indication of the degree of variation of composition. One of the difficulties, however, is that the data sets have not necessarily been derived in the same ways, different collecting methods, techniques of analysis, and methods of reporting concentrations having been used. Methods of precipitation sampling for chemical analysis and described by Junge and Gustafson (1956). Few data sets give a complete analysis and they do not necessarily refer to the same ions. Some studies focus upon rain-water while others differentiate between the contributions of particle impaction on leaf surfaces, dry fallout, rainfall solutes, and snow melt.

In a study of impaction of particles on leaf surfaces in northern England, White and Turner (1970) found P, Na, K, Ca, and Mg to be present. Similarly, Allen *et al*. (1968) demonstrated the dominance of P in rain-water together with lesser amounts of Na, K, Ca, and Mg and also some small amounts of N. Douglas (1972), in a review of the field of precipitation chemistry, cites Cl, Na, K, Mg, and Ca with reference to a discussion on the chemistry of dry air through which falling rain passes. Further examples of precipitation analyses are given by Fisher *et al*. (1968) for the Hubbard Brook

Table 4.1 *Maximum and minimum concentrations of five ions over continental USA*

Ion	Maximum (p.p.m.)	Minimum (p.p.m.)
Ca^{2+}	6.5	0.27
Na^+	8.0	0.14
K^+	0.8	0.06
SO_4^{2-}	10.8	0.69
Cl^-	8.85	0.09

Source: Leopold *et al.* (1964)

catchment in New Hampshire, USA. Here the presence of sulphate, ammonium, and nitrate are demonstrated. Sulphate and chloride are shown to be dominant constituents by Cleaves *et al.* (1970) and Likens *et al.* (1970) conclude that sulphate and hydrogen ions are the most abundant constituents of rain-water. Other discussions and data are to be found in Cryer (1986), Egner and Eriksson (1955), Fisher *et al.* (1968), Gambell and Fisher (1966), Gorham (1957, 1961), Stevenson (1968), and Whitehead and Feth (1964).

Maximum and minimum figures for five ion species in rainfall over the continental USA are given in Leopold *et al.* (1964, p. 102); see Table 4.1. The average composition of rain-water in the USSR is given by Alekin and Brazhnikova (1968); see Table 4.2.

In terms of the formulation of a budget the calculation of the annual input of ions to a drainage basin must be based on a knowledge of the amount of yearly rainfall as well as of its composition. This type of calculation has been attempted by Edwards (1973a, pp. 210–11) for two East Anglian catchments, the Yare and the Tud. The values for input expressed as metric tonnes per square kilometre are given in Table 4.3.

Of the ions mentioned so far, most are of use as plant nutrients. However, all the anions present can become involved in weathering and leaching processes as described in Section 3.4.4. This will include HCO_3^-, NO_3^-, Cl^-,

Table 4.2 *Average composition of rain-water in the USSR*

Ion	Average concentration ($mg\ l^{-1}$)
HCO_3^-	18.20
SO_4^{2-}	9.17
Cl^-	5.46
Na^+	5.12
Mg^{2+}	1.74
NO_3^-	1.70
NH_4^+	0.21

Source: Alekin and Brazhnikova (1968)

Nutrient Systems: Components

Table 4.3 *Annual input of precipitation elements to two East Anglian catchments, metric tonnes/km^2*

Ion	Input ($t\ km^{-2}\ a^{-1}$)
Na^+	1.7
K^+	0.3
Ca^{2+}	1.7
Mg^{2+}	0.2
Si	0.0
Cl^-	3.8*
HCO_3^-	0.0
PO_4^{3-}	0.0
SO_4^{2-}	1.6**

* data from Stevenson (1968).
**data from Williams and Cooke (1971)
Source: Edwards (1973).

SO_4^-, and OH^-. Moreover, although not mentioned in the Tables above, the hydrogen ion, H^+, will be present in rain-water as a consequence of the dissolution and ionization of carbon dioxide in the rain (see Section 3.4). The pH of rain-water reportedly ranges from 4 to 8 (Leopold *et al.*, 1964), while Gorham (1955) gives a range of 4 to 5.8 for the English Lake District. Atmospheric carbon dioxide equilibrates with water to yield pH values of around 5 to 6.5, as influenced by temperature and pressure. Values below this are probably due to local acidic pollution sources (see Section 4.4, below) and values above this to alkaline terrestrial dust or marine sources (the pH of sea-water being in the region of 8.2). Certainly, although rain-water will not contain so much dissolved carbon dioxide as soil water (see Section 3.4) the hydrogen ion will be present and may have a considerable effect upon soil chemical processes (see Section 4.4 and 4.5, below).

4.3. Ionic balance

An interesting point about rain-water composition is whether the weathering potential provided by anions and hydrogen ions is offset by the provision of nutrients. It is maintained by Newman *et al.* (1975) that if rain-water is to remain electrically balanced then the amount of cations present must be equal to the amount of anions present. However, even if this is so, it is likely that a net weathering potential exists in that part of the cation compliment will be made up of hydrogen ions. However, this is a difficult point to generalize on because, again, the variability of the composition of rainfall make it probable that net weathering potential and the provision of nutrients will vary from locality to locality.

4.4. Acid rain and soils

The vast amount of literature on the many causes, effects, and issues of acid rain are largely outside the scope of this book. The term 'acid rain' has become a commonly used shorthand, covering a wide range of processes and effects. The focus here is on the effects of precipitation acidity and of dry deposition of acid particulate matter on nutrient mobility in soil and vegetation systems. There are also many other sources of soil acidification in ecosystems besides atmospheric ones, including natural as well as anthropogenic sources, and these should also be discussed. Their effects on soils vary with the nature of the soil, especially in terms of the reserves of weatherable minerals which can effectively neutralize acid inputs. Acid inputs to soil and vegetation systems are important because they will influence soil acidity and the release of soil nutrients; they can also influence the leaching or nutrients from plant canopies and the losses of solutes in stream runoff.

4.4.1. Sources of acidification

In simple terms, the input of acid substances increases the mobility of elements present in compounds susceptible to acid hydrolysis by H^+, as discussed in Section 3.4.1. The principles of such processes are well established but the extent of the processes, their causes, and their effects are less well agreed upon. Moreover, the spatial and temporal effects are also highly variable, both in terms of acid input and soil response, making a consistent evaluation of the picture difficult. Sources of acidity in soil and vegetation systems can be summarized as follows (Fig. 4.2):

Atmospheric: rain, dry deposition, H^+ from H_2CO_3, H_2SO_4, HNO_3, and other acids.
Canopy leachate: leaves, stem, bark, mineral, and organic acids.
Humus decomposition: Organic acids, CO_2 ($CO_2 + H_2O = H_2CO_3 = H^+ + HCO_3^-$).
Soil respiration: CO_2 (as above).
Fertilizer ammonium: from NH_4NO_3, $NH_4 + 2O_2 = H^+ + H_2O$; Ca may also be lost as $CaNO_3$.
Biomass storage: cations removed from the soil.
Biomass harvesting: cations removed as timber or crop.

Currently, there is considerable interest in anthropogenic sources of acid inputs to soil and vegetation systems, but, as can be seen from the list above, there are many sources of acidity and it is clear that acidification of soils is not

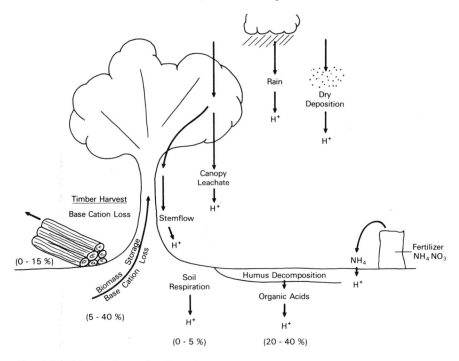

FIG. 4.2.   Mechanisms of soil acidification, involving both inputs of H $^+$ and losses of base cations. Data are for suggested ranges of contribution for forested ecosystems (northern spruce/pine on moderate/poor nutrient status soil) with the combined effects of atmospheric inputs being 25–80%. On agricultural land, fertilizer contribution may be 65–80%, soil respiration 10–20%, and acid precipitation 5–10% of acidity sources (data from Environmental Resources Limited, 1983)

only a phenomenon induced by human activity but it is also a natural tendency.

This is especially the case in regions where high rainfall facilitates the leaching of bases (see below) from the soil. However, under well-established stands of vegetation, leaching of bases from the upper horizons of soils will be offset to a certain extent by biological recycling. Here, plant roots will extract nutrients from the lower soil horizons and those nutrients not stored in standing biomass (such as tree trunks) or lost by harvesting will be returned back to the soil surface in leaf litter fall.

Biomass storage tends to decrease as an ecosystem matures. In a young stand of trees, for example, storage of nutrients in growing tissues is marked whereas in older stands, growth tends to be less vigorous and uptake tends to be more balanced by the fall of litter, branches, and trunks. Thus, the continual disturbance of ecosystems involving biomass harvest, and which leads to renewed, vigorous growth, can cause greater soil impoverishment than would be the case under a mature, stable ecosystem.

The extent of nutrient return by litter fall, as well as depending on biomass storage and harvesting, will also depend on the nature and extent of the decomposition of the litter. Complete decomposition tends to release nutrients as the decay proceeds to release the basic elements of nitrogen, carbon, hydrogen, and oxygen. This might be the case under warm, moist climatic conditions on a fertile site with an active soil biological population.

Where decomposition is discouraged by harsh climatic conditions (such as coldness or wetness) or by site infertility (or both), decomposition is incomplete and organic acids tend to be produced, often from a thick humus mat present on the soil surface. These acids are extremely effective agents of weathering (see Section 3.4.3) and they mobilize soil bases and encourage soil acidification by leaching.

In addition, in situations of high rainfall, the removal of deep rooting tree species without tree regrowth can further encourage soil acidification as their effect of recovering and recycling leached nutrients is lost. This is especially the case when trees are replaced by shallow rooting vegetation, such as grasses or heath vegetation. Thus, many acid podzolic soils are thought to have been formed in Britain after times of clearances of primary forest by prehistoric man. Similar impoverishment also appears to follow the clearance of tropical rain forests.

Replacement of deciduous tree species by conifers can also increase soil acidification because conifer needles tend to be more acidic than broadleaves (Fig. 4.3), though there is an overlap in the ranges of acidities for the two tree types and the extent of the effect also depends on the reserves of weatherable minerals which are present in the soil and which can offset the effects of acid inputs, as discussed below.

Other activities can also be important, such as drainage. This may encourage decomposition processes by increasing biological activity but it can also lead to the oxidation of acid minerals, forming acids in solution (such as pyrite oxidation which leads to the formation of sulphuric acid).

Thus, there are many direct and indirect ways in which human activity can affect soil acidification, and the longer term cycling effects can be as important as the more immediate effects of rainfall acidity and ammonia-based fertilizer addition. It should also be stressed, however, that not all human activity involves acidification because agricultural practices, such as liming, act to decrease acidity, as does limestone quarrying, where calcareous dust may be blown from the site of quarry operations, acting to decrease acidity in the local environment. Acidification can therefore involve several sources, both natural and human-induced and it may also be offset by activities which decrease acidity.

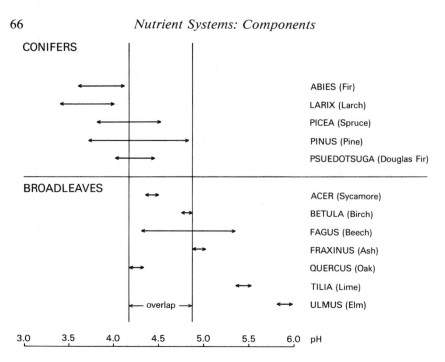

FIG. 4.3.   Ranges of acidity of aqueous extracts of conifers and broad-leaved trees (drawn from data in Handley, 1954)

4.4.2. Soil acidification

Whatever the source of acidification, the effect it has on soils will depend upon the soil type. Simply expressed, if a soil is already very acid, a further acid input will have little additional effect. In alkaline, calcareous soils, there will also be little effect, but in this case it is because there are enough alkaline minerals present in the soil which can break down to offset acid inputs by weathering reactions. It is, in fact, the weakly acid or neutral soils which are the most prone to acidification because they have only low reserves of weatherable minerals which can offset acid inputs.

Buffer capacity.   The ability of a soil to resist changes in acidity is termed the *buffer capacity* of the soil. Overall buffer capacity consists of several components which vary in their availability in a similar manner to the availability of nutrients shown in Fig. 2.3. They range from the immediately available reactants in the soil solution, reactants present on the exchange complexes of clays and organic matter and the less readily available reactants present within minerals and organic matter and which are only released by weathering or decomposition. The reactants which offset acid inputs are the *bases*, defining a base as a proton (H^+) acceptor (such as CO_3^{2-} derived from the dissociation of $CaCO_3$ in water, and which then combines with H^+ to form hydrogen carbonate, HCO_3^-. Bases present in solution can readily offset acid inputs in solution and once these have been used up, exchangeable bases will then act to neutralize input acids. Further supplies of bases can only then be provided by the breakdown of solid phases within the soil. This progression of availability gives rise to the notion of the readily available *active* and the less readily available *reserve* buffer capacities.

In calcareous soils, such as those with abundant limestone particles, there is a highly available source of bases in the soil solution, and reserve buffer capacity is made available if the acidity of the soil solution drops, solubilizing the limestone particles. Acid soils, on the other hand, have their exchange complexes dominated by hydrogen ions and aluminium (Fig. 4.4). Some buffer capacity to offset acidity can be provided by the breakdown of clay particles themselves, but in sandy, quartz-based soils, reserve buffer capacity is extremely limited. These soils are almost certainly going to be acid in any case because of natural leaching processes and any additional acidity, say from acid atmospheric deposition or from conifer needles, is going to have very little additional effect.

Thus, soils in the range pH 4–5 will be little changed by acid inputs as they are acid already. Conversely, soils with pH values above 7 are only thus because they have a high buffer capacity which keeps them at that level by the weathering of base-rich minerals. It is the soils in the pH range of 5–7 which are the most vulnerable to acidification because the reserve buffer capacities are only limited. If they were any higher, the soil would have a higher pH and

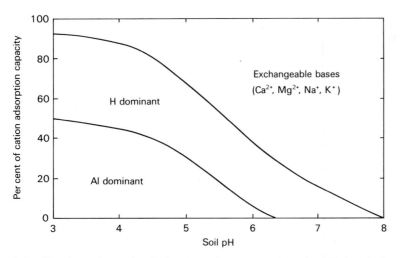

FIG. 4.4. Dominant ions adsorbed onto exchange complexes in soils in relation to pH. Base cations decrease in importance with increasing acidity, to be replaced by hydrogen ions and then aluminium (adapted from Brady, 1974)

be resistant to pH change; if they were any lower, the soils would be acid anyway. The concern about the acidification of soils therefore rests with these slightly acid soils of limited buffer capacity. They can be readily impoverished by the planting of acid needle-bearing conifers or by atmospheric acidic deposition.

Buffer capacity can be readily assessed by the use of an acid-base titration. Here, dilute acid is added drop by drop to a soil paste and any pH changes measured during the course of the addition. Soils with high buffer capacities will be resistant to pH change and soils with low buffer capacities will rapidly give acid readings. In this way, the susceptibility of different soils can be compared. An example is given in Fig. 4.5 where contrasting soils have been treated with acid; notice that the acid peat soils show little change from their already acid values, soils in the pH 5–6 range are at first resistant to acidification, but thereafter drop rapidly, and that limestone soils show considerable prolonged resistance of pH change.

It can be seen that soils can vary markedly in their susceptibility to pH change. Therefore the effects on the nutrient status of soil and vegetation systems can also vary markedly. Nutrient mobility is discussed further in Section 4.4.4 but an additional factor which must be considered is the way in which soil hydrological processes influence nutrient mobility and runoff acidity.

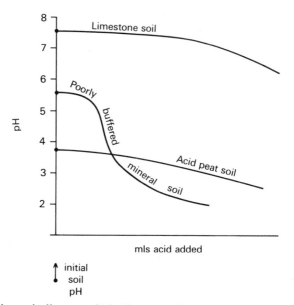

FIG. 4.5. Schematic diagram of 3 buffer curves for contrasting soils. Limestone-rich soils have a high buffer capacity and pH remains high; the buffer capacity of mineral soils with low reserves of weatherable minerals is soon used up and acid peat soils show little change from their already acid values

4.4.3. Runoff acidification

The effects of the acidification of soils, by whatever source, illustrate the closeness of the links between the soil and vegetation systems components which interact through the medium of water. As has been emphasized before, the system will react as a whole if stressed at a particular point. Increased acidity represents such a stress to the system, with considerable effects on runoff water quality, as discussed below, and on nutrient mobility, as discussed in Section 4.4.4.

It is clear that catchment processes, especially the transfer of water over and through the soil, are of crucial importance in determining the acidification of runoff. Research work has indicated the importance of soil hydrochemical processes in determining stream-water quality (e.g. that by Bache, 1983, 1984; Miller, 1985; Shaffer and Galloway, 1982; Cosby, Hornberger, and Galloway, 1985). These processes can modify the effects of acidity inputs to the soil from rain, dry deposition, and vegetation. In addition, the inputs themselves are influenced temporally by atmospheric source area (Fowler, Cape, and Leith, 1985), and spatially by land use cover, especially coniferous forests (Hornung, 1985).

Reinforcement of these points of view comes from a review assessment of acid precipitation, and the problems of its effects, by the Director of the Royal Society's Surface Water Acidification Programme, Sir John Mason, who writes: 'Of central importance is the extent to which the chemistry of the rain and snow water is modified as it penetrates through soil and rocks before arriving in the streams and lakes' (Mason, 1986). Several agencies and mechanisms of modification are then listed, including soil neutralizing capacity, calcium and aluminium contents, oxidation of sulphur and ammonium, and the effects of conifers in acidifying soil. Mason concludes: 'The relative importance of these various agencies and mechanisms will vary a great deal between different catchments. In order to determine their contributions in quantitative terms in a number of representative catchments . . . will require a major effort involving detailed and sustained investigation of the whole chain of events that intervene between power station chimney and the gills of the fish.'

Soil hydrological processes.   Soils vary considerably in their ability to neutralize atmospheric and vegetation input acidity, and this can be predicted through measurements of their buffer capacity. Such assessments indicate the *potential* of the soil to neutralize acidic inputs. But, notwithstanding the usefulness of such buffer capacity assessments, (Tabatabai, 1985), the *actual* amelioration of acidity will depend upon the way in which runoff processes influence the *possibility* of input rain-water mixing with the soil solutions and its degree of contact with the solid phase of the soil, as already realized by Voigt (1980), Bache (1983, 1984) and Miller (1985).

Rapid flow through structural cracks, root channels, and other macro-pores will minimize the opportunities for mixing and contact. Similarly, the occurrence of saturation overland flow, maximized on wet soils of low soil infiltration capacity, will limit the opportunity for rain-water to interact with the soil medium. In addition, return flow, while this might have become equilibriated with the mineral phases of the soil, can lose bases by exchange processes as it reappears through organic soil to flow overland at slope foot sites.

Thus hillslope hydrological processes are significant, both on the upslope areas and also on the slope foot and riparian zones. It can be predicted that areas with either rapid soil throughflow, or high amounts of overland flow, will contribute the least modified (and, by inference, the most acidic) runoff water to streams. The crucial factors are thus the ways in which soil charac-teristics influence the occurrence of overland flow and also the extent and rate of soil throughflow.

Interactions between soil hydrological and chemical processes.   In addition to the consideration of the attenuation of acid inputs by soil hydrological processes, there is the allied point that atmospheric deposition is not the only source of runoff acidity. Additional sources of acidity can come from the surface organic soil horizons. This is especially the case on acid, peat soils. If such soils have a limited infiltration capacity, not only will there be a limited opportunity for chemical reaction with the mineral phase of the sub-surface soil horizons, but acidity may well be actually increased over that derived from the atmospheric input by the uptake of organic acids from soil surface organic matter.

Specifically, if a situation is found where acidic stream runoff is known to occur, then the presence of acid waters in streams must represent either:

1. A situation where there was adequate buffering capacity in the soil but there was the rapid transfer of acid rain-water (or rain-water acidified during stemflow or throughfall from vegetation) to the stream by either:
    (a) rapid overland flow because of saturated soil conditions/low soil infiltration rates, or
    (b) rapid, preferential flow in soils via structural cracks or other macropores under conditions of high rainfall intensity,
    both situations giving rise to limited neutralizing contact with the soil medium.
2. A situation where adequate buffer capacity exists in the mineral soil, but where there was a marked uptake of acidity from the organic layers of the soil, again followed by lack of neutralization by the soil medium, as given in 1(*a*) and (*b*) above.
3. A situation where there was a lack of buffer capacity in the soil, so that

even if flow was not diverted overland and even if it flowed slowly and uniformly through the soil, the soil had no buffering potential to offset input acidity from whatever source.

4. A situation where there was rapid overland flow, or rapid preferential throughflow, and no buffer capacity in the soil.

Of the possibilities outlined above, Bache (1983) suggests that virtually all soils are able to neutralize even heavily polluted rain and thus the possibilities involving a lack of buffer capacity look to be the least likely. This therefore stresses the importance of soil hydrological processes in controlling the acidity of runoff. Nevertheless, the lack of neutralizing potential cannot be discounted, particularly in peat soils and acid, sandy, quartz-dominated soils.

It can thus be seen that the important factors are the ways in which acidity sources interact with soil hydrological and chemical processes (Fig. 4.6). These combine to control the effects of acidic inputs on nutrient mobilization, specifically in terms of whether acidity is attentuated in the upper layers of the soil or whether nutrients are mobilized within the soil and are thus lost to lower soil layers or to stream runoff.

4.4.4. Nutrient mobility

It is clear from the foregoing discussion that if atmospheric sources of acidity are not attenuated in the soil, because of low buffer capacity or hydrological by-passing of the soil mineral phase, then the acidity of runoff waters will be increased. However, if such acidity is attenuated in the soil by the available buffer capacity, the processes involved will be either the exchange of hydrogen ions for other cations adsorbed on clay–humus complexes or the

(a)   (b)

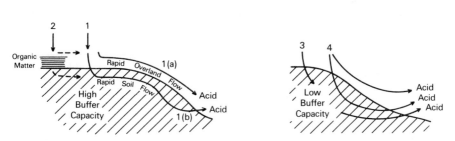

FIG. 4.6. Schematic representation of causes of acid runoff on a hillslope with (a) high soil buffer capacity; and (b) low soil buffer capacity. In (b), runoff is acid whatever water flow rate or route occurs, in (a), acid runoff occurs in relation to uptake from organic matter and/or rapid overland flow or throughflow

hydrolysis of mineral phases. In either case, there will be a release of cations into the soil solution where there will be a tendency for them to be leached from the soil into streams. Thus, increased acidity in the system inputs can result in increased cation concentrations in the system outputs. Therefore, while attenuation of acidity in soils can be beneficial to runoff water quality, in terms of the lack of acidification of streams, in can be detrimental to the soil nutrient store, in terms of increased mobilization of cations through increased hydrolysis and replacement of cations on exchange complexes.

During slight acidification, the principle cations released will tend to be calcium, magnesium, sodium, and potassium, especially those adsorbed on the exchange sites of clays. However, marked acidification can lead to the hydrolysis and breakdown of the clay minerals themselves and other minerals, with a release of the their constituents into solution, especially aluminium. Thus, Bache (1985) describes the breakdown of potash felspars (orthoclase) by hydrolysis:

$$2KAlSi_3O_8 + 2H^+ + 9H_2O \rightarrow Al_2(OH)_4Si_2O_5 + 4Si(OH)_4 + 2K^+$$
$$(4.1.)$$

This type of reaction leaves residual clay minerals and produces soluble silica and soluble cations (specifically, K^+, above), the latter normally being leached as their bicarbonates. As acidification proceeds, the aluminium is released by the breakdown of clays in a process which becomes significant at soil pH values below 5.5:

$$3H^+ + Al(OH)_3 \rightarrow Al^{3+} + 3H_2O \qquad (4.2)$$

The aluminium may be present in the form of a soluble complex with silica or in an organically complexed form. Thus, at an early stage, plant nutrients may be lost and at a more advanced stage, aluminium can be released into the soil solution, where it may have a deleterious effect on plant growth, or into streams, where it can become toxic to fish and decrease the value of water supplies for drinking purposes.

Since many of these chemical elements are released into mobile soil solutions and thence into runoff waters, many studies have focused on the study of runoff water quality in relation to atmospheric acidic inputs. Thus, Cronan and Schofield (1979) chart the changes in acidity and chemical element release as water moves from precipitation to vegetation throughfall, soil percolate and spring water in a non-calcareous, podzolic soil catchment in the USA (Table 4.4).

It is clear that pH attenuates through the system, as shown more clearly by the H^+ values, and that cation release is progressive, especially for aluminium.

Similarly, Cosby *et al.* (1985a) assert that ' . . . surface water chemistry is determined by reactions that occur in the soils and rocks within a catchment'.

Table 4.4 *Changes in water quality showing pH attenuation and cation release, data in microequivalents per litre*

Sample	pH	H$^+$	Ca^{2+}	Mg^{2+}	K$^+$	Na$^+$	Al*
Bulk precipitation	4.08	83	9	3	3	4	n.d.
Throughfall	4.02	95	36	14	37	3	5
Percolate	4.02	91	25	15	16	7	54
Springs	4.66	22	26	19	10	13	67

*Aluminium as the summed total of Al^{3+}, AlOH^{2+} and Al(OH)$_2{}^+$.

n.d. = no data available.

Source: Cronan and Schofield (1979)

They also show that aluminium initially exchanges for base cations held on adsorption sites:

$$Al^{3+}{}_{(a)} + 3BC_{(x)} \rightarrow Al_{3(x)} + 3BC^+_{(a)} \qquad (4.3)$$

where $_{(x)}$ denotes an adsorbed phase, $_{(a)}$ an aqueous phase in solution and BC$^+$ is a base cation. Thus, base cations move into solution while the aluminium ions are adsorbed because the cation exchange sites on the soil matrix have a higher affinity for trivalent ions, such as aluminium, than for the divalent or monovalent base cations. Eventually, if the exchangeable base cations in the soil become depleted by this process, further aluminium cannot be exchanged and so the Al^{3+} concentration in the soil water and runoff water becomes higher and the base cation levels become lower. The process can only be offset by the weathering of minerals and thus the reserve buffer capacity becomes a crucial factor in controlling acidity and aluminium levels and thus also in controlling nutrient mobility. In addition, the leaching of the base cations is effected by the presence of strong acid anions, such as SO$_4^-$, NO$_3^-$ and Cl$^-$. The protons (H$^+$) from acids become adsorbed onto clay sites and the acid radicals (the anions) combine with the base cations to preserve electrical neutrality in solution.

In essence, these processes are no more than accelerated forms of normal chemical weathering by hydrolysis, as emphasized by Webb (1980). However, the increased mobility of the chemical elements and the increasing saturation of exchange complexes by aluminium means that the elements are not held in the soil by adsorption, but are progressively lost to runoff waters.

Thus, soil processes can have a marked effect on runoff water quality, and, working on the impact of acid precipitation on headwater streams in Virginia, USA, Shaffer and Galloway (1982) compare input water chemistry with soil chemistry and stream output chemistry. They show that acidic cations and anions are accumulating in the catchment studied, including H$^+$, NH$_4^+$, NO$_3^-$, and SO$_4^-$, while there is a net export of the base cations K$^+$ and

Mg^{2+}, together with HCO_3^- and silica. Such observations act to confirm the effectiveness of the processes proposed above by such workers as Bache, and Cosby *et al.*

The soil chemical processes not only include the chemical reactions described above, but may also include the adsorption of nutrient cations present in precipitation on soil humus layers (Låg, 1976). This process can act to retain nutrients in the soil and vegetation system and thus may help to partially offset the mobilization processes already discussed.

Modelling the processes involved has been attempted by Galloway *et al.* (1983) and by Cosby *et al.* (1985b). Taking the specific case of acidification by sulphuric acid deposition, the former group of authors suggest that there are four stages in acidification and nutrient mobilization:

STAGE 1. Systems are not receiving acid deposition. Losses of base cations are adjusted to input acidity, weathering rates, and biogeochemical cycling.

STAGE 2. Systems are receiving acid deposition but the soil is not yet saturated with SO_4^{2-} but the acidification is having the effect of increasing the amounts of base cations lost in solution.

STAGE 3. Systems are receiving acid deposition and the soil is saturated with SO_4^{2-} and the depletion of base cations has resulted in a decreased output of base cations in solution.

STAGE 4. As above, but base cation output has stabilized at a new, lower level.

This model is illustrated in Fig. 4.7.

In a more developed model, Cosby *et al.* (1985b) combine both chemical and hydrological factors, as already discussed in Section 4.4.3. They consider the processes of soil cation exchange, aluminium dissolution, and carbon dioxide dissolution together with spatially distributed catchment hydrological processes. The model assumes a rapid equilibriation time of the chemical processes and twenty-four chemical equations are used to predict the aqueous chemistry of runoff waters. The basic parameters include aluminium mobility and base saturation of soils as well as acidification processes. Such models are liable to be important in future progress in understanding acidification trends in soil and vegetation systems, using input acidity and output water quality as major parameters.

In Section 4.4, we have briefly reviewed selected aspects of the 'acid rain' issue, especially as it affects nutrient mobility in soil and vegetation systems. It is clear that acidification sources are numerous, and not just anthropogenic. It is also clear that the affects vary according to soil chemical and hydrological characteristics. Nevertheless, the overall indication is that acidification initially increases the mobility of nutrients, and ultimately leads

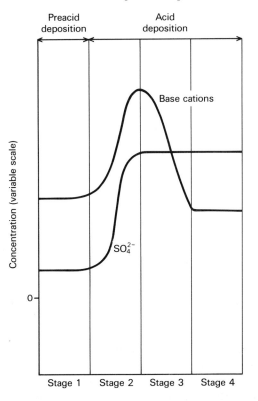

FIG. 4.7. Four stages of acidification, with the response of base cation release and sulphate concentrations in soil, for details of STAGES 1–4, see text (modified from Galloway *et al.*, 1983)

to the impoverishment of the nutrient store in soil and vegetation systems. This is not to deny the importance of precipitation elements as inputs to soil and vegetation systems, a short discussion of which forms the last Section of this Chapter.

4.5. The role of precipitation elements in nutrient systems

The deposition of atmospheric nutrients is highly variable in time and space and their significance to the nutrient supply of soil and vegetation systems also varies considerably with soil and vegetation type. The inputs act to offset losses from leaching and also from timber harvesting. They can also have considerable impact upon weathering processes, the acquisition of base cations acting to slow down weathering processes while the acquisition of H^+ and acidic anions acts to increase weathering rates, as shown above (Section 4.4) and as discussed further below.

4.5.1. Nutrient inputs

Precipitation elements are a major source of nutrients in areas where the rock substrate provides little in the way of nutrients, especially in hard rock upland areas (White *et al.*, 1971) or in moorland areas (Crisp, 1966). In such infertile situations ion exchange process appear to be important during the percolation of the rainfall through the raw humus (Låg, 1968). However, in other situations (for example, in the work of Feth *et al.*, 1964) where lithospheric supply of nutrients is adequate then precipitation inputs are less important, but can still be significant. In addition, under tree canopies, foliar leaching can augment the supply of precipitation elements, as has been suggested by the work of Carlisle and White (1966), Attiwell (1966), and Madgwick and Ovington (1959). The processes involved here are the removal of plant nutrients from the foliage as well as the leaching-out of organic substances with their ensuing deposition on the soil.

The general role of precipitation elements in ecosystem processes is discussed by Eriksson (1960), Zverev (1968), Swank and Henderson (1976) and Swank (1984). Swank (1984) shows that nitrogen in bulk precipitation (wetfall + dryfall) is equivalent to at least 70 per cent of the nitrogen incorporated annually in above-ground woody tissues of some temperate hardwood forests. In addition, atmospheric sources of calcium and potassium supply between 20 and 40 per cent of the nutrients taken up by increments of woody tissues (Table 4.5).

Such precipitation inputs are augmented by canopy leaching during throughfall and stemflow, often by as much as 20 to 100 per cent. Comparison of throughfall loadings (kg ha^{-1} a^{-1}) with precipitation loadings always yields a positive ratio, showing that throughfall inputs were greater than precipitation inputs. Data for inputs to the forest floor are quoted to range from a throughfall/precipitation ration of 1.2 to 1.7 for nitrogen, 6 to 33 for potassium, 1.6 to 3.1 for calcium, and from 2.2 to 40 for phosphorus,

Table 4.5 *Comparison of selected nutrients in bulk precipitation (P) and stored in annual woody increments (I) for two hardwood forests in the eastern USA. Data are in kg ha^{-1} a^{-1}*

Location	Total N		K		Ca	
	P	I	P	I	P	I
Walker Branch, Tennessee	13.0	15.0	3.0	8.0	12.0	31.0
Coweeta, North Carolina	8.8	13.0	2.1	13.0	4.8	23.0

Source: Swank (1984)

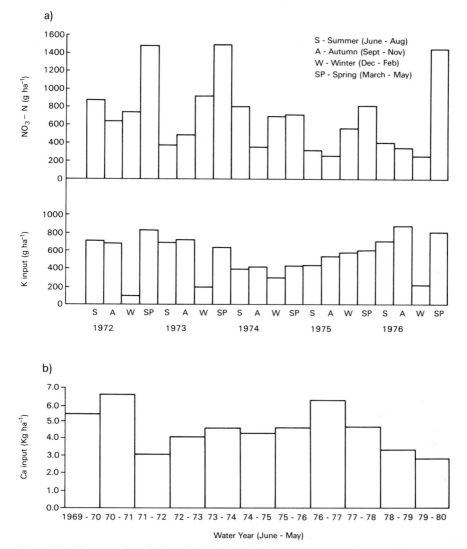

FIG. 4.8. Temporal variations in the chemical inputs from bulk precipitation at Coweeta Hydrologic Laboratory, North Carolina, USA: (a) potassium and nitrate–nitrogen, showing seasonal variations from 1972–76; (b) calcium inputs for 1969–80, showing variations in mean annual weighted concentration (modified from Swank, 1984)

implying considerable enrichment of all four elements, especially potassium and phosphorus.

There are considerable variations in precipitation inputs from month to month and from year to year (Fig. 4.8). This means that the evaluation of the role of inputs has to proceed with caution and any data base must be a long

Table 4.6 *Nitrogen balances: comparison of precipitation inputs and losses consequent upon tree harvesting*

(a) Oak–hickory forest, whole tree harvest, 90 year rotation

	kg ha^{-1} a^{-1}
Nitrogen Losses	
Leaching to runoff	0.5
Harvest	5.3
Denitrification	15.5
Nitrogen Gains	
Fixation	10.9
Precipitation	8.8

(b) Coweeta, North Carolina, inputs and outputs after clear felling*

Bulk Precipitation	8.82
Leaching to runoff:	
Year 1 after clear cut	9.70
Year 2 after clear cut	6.87

*NO_3-N, NH_4-N and Organic N (dissolved + particulate + sediment)

Source: Swank (1984)

term one before any realistic estimates can be made. In such endeavours, the use of ion-exchange resins at the base of precipitation collectors may be useful, as suggested by Crabtree and Trudgill (1981). Such modified collectors can give useful long term data about bulk precipitation integrated over long time periods.

One important point which has emerged from the study of precipitation inputs is that it has become apparent that such inputs can substantially offset losses from harvesting also those from the nutrient leaching which follows the disturbance of clearcutting (Table 4.6). Nitrogen fixation and denitrification remain the dominant processes involved in nitrogen supply but for base cation nutrients not accreted and lost by these processes, the suggestion is that a consideration of precipitation inputs is vital to the understanding of the nutrient balance of disturbed soil and vegetation systems.

Thus, while precipitation inputs represent the major supply of nutrients for nutrient poor and raised peat soils, they are also important on more nutrient rich soils. The total inputs may not be as much as the reserves in the weatherable soil minerals, but they represent significant additions and these can be important in offsetting harvesting and leaching losses.

4.5.2. Weathering

In the context of acid rain, it was emphasized how acid inputs can act to release nutrients from exchange sites and minerals, eventually leading to nutrient impoverishment in extreme cases, especially on nutrient-poor soil parent materials. In less extreme cases—and not just in the context of the acid rain issue—it should be emphasized than rainfall chemical composition normally exerts an influence on the weathering of soil minerals and thus on the release of nutrients. This is both in terms of acid hydrolysis and also the leaching of base cations in combination with precipitation anions.

In the context of the hydrolysis of minerals and replacement of adsorbed base cations by H^+, Fisher *et al.* (1968) suggested that the annual input of hydrogen ions in rainfall is equivalent to the total discharge of base metal ions in stream output of the Hubbard Brook catchment. This implies that rain-water is a major source of reactants provided for weathering processes and that weathering is closely related to rainfall chemistry, although the inter-mediate soil processes are not necessarily clear.

The review of Carroll (1962) suggests that rainfall amount and chemistry combine to influence the type of clay produced by weathering reactions in the soil. It is suggested that montmorillonite is produced, with adsorbed Na^+, Mg^{2+}, and Ca^{2+} present, under favourable circumstances of low rainfall and high nutrient supply. However, as rainfall amount and acidity increase, calcium is lost, giving Mg^{2+}, Na^+, and H^+ on the exchange complex until eventually kaolinitic clay, with adsorbed H^+, is left as a residue. Further-more, Carroll suggests that the acquisition of cations from rain-water by clay minerals may actually slow down the rate of chemical weathering in the soil. Rainfall chemistry and amount, while variable, can thus be a vital influence upon soil chemical processes.

In summary, rainfall chemistry is an important topic for consideration in terms of its influence upon the soil nutrient store. In the case of infertile sub-strates it represents a major source of nutrient supply. However, where potentially weatherable minerals occur in the soil the amount of rainfall and the balance of the content of nutrient cations (such as Ca^{2+}, Mg^{2+}, and K^+) relative to the presence of substances promoting weathering reactions (H^+ and anions) will be a crucial factor influencing the progression of weathering in soils and the acidity and nutrient status of soils.

Further reading

Rain as a geologic agent; rain-water supply of nutrients to ecosystems; rain-water chemistry. (Carroll, 1962; Cryer, 1986; Fisher *et al.*, 1968; Stevenson, 1968.)

5 The Leaching Output

The man to put your money on is not the man who wants to make 'a survey' or 'a more detailed study' but the man with the notebook, the man with the alternative hypotheses and the crucial experiments.

JOHN R. PLATT

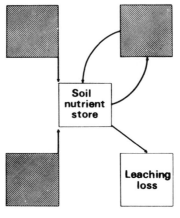

FIG. 5.1.

5.1. Introduction

Water flow in soils is an important factor in soil nutrient systems since it brings the reactants to the sites of weathering reactions and it will remove the products from the weathering sites (as has been outlined in Sections 3.3.1 and 3.3.3). Products removed from the sites of weathering may be redeposited at the base of the soil profile or they may be lost altogether from the soil as a leaching output to streams or groundwater.

This chapter does not consider the removal of weathering products by plant uptake, but rather discusses inorganic chemical and hydrological processes involved in leaching loss. The relative importance of losses by plant uptake and by leaching are discussed in Chapter 7.

The redistribution of chemical elements within the soil profile may operate in a movement which is separate from the movement of chemical elements through the soil mass and out into fluvial (or groundwater) systems. It is only the more soluble elements which will be lost in the latter, leaving the less soluble elements in the soil profile. Solubility, however, depends upon the environment of weathering, as discussed in Chapter 3. For example, it appears that iron chelates can be deposited at the base of the soil profile because the chelates become less soluble as they pick up iron (see Section 3.4.3). However, this is not necessarily the fate of all iron in a soil profile as reddish-coloured iron-rich compounds may often be found draining into streams from waterlogged soils. Clearly, the mobility of individual elements depends upon weathering agency and environment and careful consideration should be given to solute mobility as water passes from one environment to another.

5.2. The soil profile

Redeposition in the soil profile can depend upon a number of factors and they do not always include the downward movement of leachates. In some cases, where the base of the soil is waterlogged and anaerobic, but where the upper part of the soil is freely drained and aerobic, deposition can occur at the top of the water table by oxidation and the formation of less soluble substances. Below the water table, the substances may be reduced and soluble, but above it oxidized and precipitated. This may be a factor in iron deposition in soils where necessary conditions obtain.

Redeposition can occur in most soils for a variety of biological and chemical reasons. Iron compounds may become polymerized and less mobile, as well as less soluble as they pick up iron. Biological action may act to break down chelates and release the contained elements. However, pH change remains the most likely reason for deposition to occur, coupled with climatic factors like seasonal drying.

In soils which are not always moist chemical saturation and precipitation will occur as the soil dries out. This will be especially true of seasonally dry soils and soils occurring in other areas where rainfall events are widely spaced in time. Water percolation down through the soil will initially dissolve solutes but as the water moves down to the lower, less moist, layers of the soil it may simply cease to flow. If the input supply is limited then the water will disperse into the soil until surface tension is too great and the water will remain static in thin films rather than moving under gravity. Though thin films are capable of moving, as unsaturated flow, eventually as the input supply ceases the gradient of differential water content will cease to be sufficient to cause water to flow. Static water will become chemically saturated and, especially if it already has a high solute load or if the water begins to evaporate, or both, precipitation may occur. In carbonate-rich soils where calcium carbonate will tend to be held in solution by combination with carbon dioxide, then degassing of carbon dioxide in aerated soils may be a factor in carbonate precipitation. For example, soils on the Jurassic Cotswold Limestones often show carbonate deposition at depth and this may be due to a combination of the factors described above. Moreover, movement of solutes may occur only over a short vertical distance downwards if a single rainfall even occurs in a period of otherwise dry weather.

In climates where dry conditions follow wet conditions, chemical precipitation will be encouraged as the soil dries out and thin water films become chemically saturated. If a standing water table exists at a shallow depth in a soil, water may also move upwards in dry conditions, depositing solutes on the surface of the soil as the water evaporates.

Rainfall frequency and amount are thus crucial factors in determining the position of deposition in a soil profile, and indeed Chorley (1969) shows a diagram which correlates the depth of calcium carbonate deposition in a soil with the amount of annual rainfall an area receives (Fig. 5.2).

5.3. Leaching losses

In areas of high rainfall, many solutes will be lost from the soil and they will be present in groundwater and stream drainage waters. The removal of solutes often represents a dominant denudation process in temperate pluvial areas (Carson and Kirkby, 1972, Ch. 9). This process of leaching loss to streams is discussed in the remainder of this chapter, given the qualification that the appearance of a solute in soil waters draining into a stream may not necessarily be indicative of the mobilization processes which are occurring in the upper part of the soil profile. Some of the mobilized elements may be reprecipitated at the base of the soil profile. Thus losses to streams may be viewed as the net loss of upper profile mobilization minus lower profile deposition.

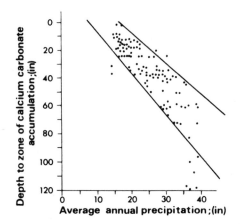

FIG. 5.2.   Depth of calcium carbonate accumulation in soils as a function of average annual precipitation (from Chorley, 1969)

Studies of stream chemistry and solutes in drainage waters from soil are found in the general reviews of Alekin and Brazhnikova (1968), Douglas (1972), and Owens (1970). Works in an agricultural context include Williams (1970) and Symposium (1973), while Edwards (1973a and b) discusses the topics in lowland agricultural catchments. Observations in upland areas are described by Lewin *et al.* (1975), Verry (1975), and Zeman and Slaymaker (1975). The use of drainage water analysis in pedogenetic studies is discussed for a chalk area by Perrin (1965).

Three important processes must be considered in order to understand leaching losses and the provision of solute to streams. These are (1) chemical reaction rate, (2) the rate of water flow in soils, and (3) the route which water takes on hillslopes. These are discussed in turn. The significance of rainfall regime is discussed in Section 5.4.

5.3.1. Chemical reaction rates

In order to understand leaching processes it will first be necessary to describe the patterns of dissolution in static waters, in other words in closed systems. This has already been discussed in Section 5.2 but the important factor in a leaching context is the pattern of solute concentration over time. In a closed system the pattern will, as a general rule, be as illustrated in Fig. 5.3. The solute concentration rises rapidly at first but then, as the products accumulate in the solvent, the rate of dissolution decreases progressively until there is no further change. This level (t_E on Fig. 5.3) is chemical equilibrium and the concentration of the solute at equilibrium (E) defines the solubility of the solute in the solvent in question.

Closed systems, with static water present, are liable to be found in soil pores between rainfall events. Considering the intermittent nature of rainfall events (see Section 5.4) it can be seen that it is likely that periods of high flow during and immediately following rainfall will be succeeded by periods where relatively static water exists, or at least only very slow flow occurs.

An important point to consider is whether the water is resident in soil pores long enough for equilibration to occur. Simply, if residence time is longer

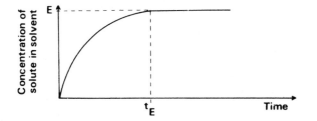

FIG. 5.3. A simple model of the general pattern of solute concentration in a solvent over time during dissolution

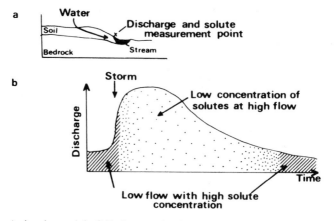

Fig. 5.4.　A simple model of discharge of a slope foot drainage output, showing dilution of solute concentration at peak storm flow (a) Section of slope showing location of discharge outlet, x. (b) Time-discharge relationship and solute pattern

than t_E on Fig. 5.3 then the dissolution can proceed until E is reached. If, however, water residence time is less than t_E the water will be undersaturated with respect to the solute. In this way, waters draining through soils at the peak and rain storms, and just after the peak, will tend to be dilute; drainage waters at other times will tend to be chemically saturated. A simple model is presented in Fig. 5.4.

Although solute concentration will tend to drop by dilution during times of high flow, this does not necessarily mean that the total amount of solutes being removed is necessarily less. The volumes of water may be so great at high flows that the total bulk of solutes removed may be equal to, or greater than, that removed per unit of time at low flows. For example, seepage water from a slope foot discharge site may have a discharge of, say, 1 l per day. If each litre of water contains 100 mg of solute then the total amount removed per day is 100 mg. If the flow rate were to increase to 10 l per day because of heavy rain, the solute load could drop to 10 mg per litre. However, since 10 l are passing each day, and each litre contains 10 mg, then exactly the same amount as before, 100 mg, is being removed each day. Care must therefore be taken that considerations of solute load take into account temporal patterns of variations in both solute concentrations and water discharge.

Studies of chemical reaction rates which can be used in the context of water flow systems can be found in many works dealing with the chemistry of natural water systems. Useful sources are those of Deju (1971), Deju and Bhappu (1965, 1966), Gould (1971), Polzer (1967), and Wiklander and Andersson (1972). Morgan (1967) provides a useful discussion on the limitations of the application of chemical thermodynamics to natural water systems. One of the points that can be derived from these works is that while

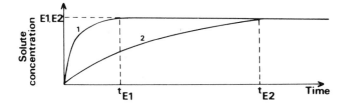

FIG. 5.5. Dissolution curves for two minerals, 1 and 2, having comparable solubility, E, but different dissolution rates

the patterns shown in Fig. 5.3 may be observed for solutes in general, the time scales involved will be different for each solute. Thus on Fig. 5.3 the horizontal scale and the value of t_E may be a matter of minutes, hours, days, weeks, or years, depending upon the solutes and solvents involved. Moreover, if complex minerals are involved, the individual constituents may be dissolving at different rates (see Section 3.3.2) even though they may have comparable final solubilities, as shown in Fig. 5.5.

Thus a residence time of water of t_{E1} will permit the dissolution of mineral 1 to equilibrium but not of mineral 2. Intermediate residence times will still occasion the preferential dissolution of mineral 1. It is not until residence times greater that t_{E2} that both minerals will dissolve equally.

There are many variations on this theme. For example, two minerals may have comparable dissolution rates but different equilibrium levels (Fig. 5.6). Here the t_E values are almost identical. Only long residence times will clearly bring out the differences in equilibrium values $E1$ and $E2$. The shorter the residence time the closer the solute concentrations will be. In addition, there is the case where both dissolution rate and solubility level of mineral 2 are lower than mineral 1. Preferential dissolution will then be encouraged by long residence times (Fig. 5.7). In this case the relative positions of the t_E values are less important because preferential solution of mineral 1 will always occur. Examples of different reaction rates for naturally occurring minerals are given by Deju and Bhappu (1965) and these are shown in Fig. 5.8.

An interesting case is where the solution curves cross over (Fig. 5.9). Here

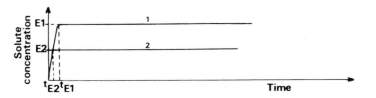

FIG. 5.6. Dissolution curves for minerals with similar dissolution rates but different equilibrium levels

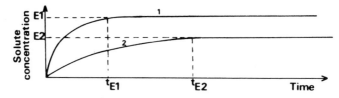

FIG. 5.7. Dissolution curves for two minerals where mineral 2 has both a lower dissolution rate and a lower equilibrium level than mineral 1

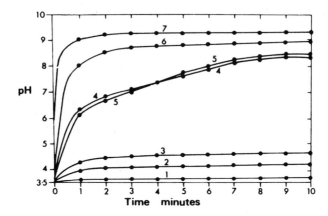

FIG. 5.8. Dissolution curves for minerals dissolving in acidified water. 1: quartz, 2: albite, 3: beryl, 4: actiniolite (amphibole), 5: hornblende, 6: enstatite–augite (pyroxene), 7: forsterite (olivine) (from Deju and Bhappu, 1965)

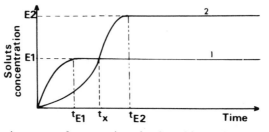

FIG. 5.9. Dissolution curves for two minerals where 2 has a slower initial dissolution rate but a higher final solubility than 1

mineral 2 has a higher E value but a much longer t_E value. In this case, preferential solubility will be closely dependent upon water residence time. For short residence times, up to time t_x, mineral 1 will be dissolved preferentially (despite its eventual lower equilibrium level) but for long residence times mineral 2 will be dissolved preferentially. An example of this is given in Fig. 5.10.

The pattern can be complex, as indicated above. Moreover, the final values

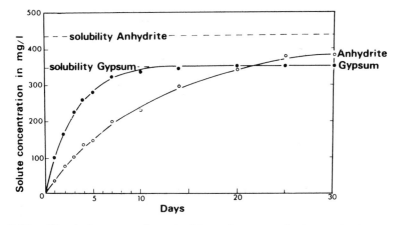

FIG. 5.10. Dissolution curves for anhydrite and gypsum, the former having a lower dissolution rate but a higher eventual solubility than the latter (from Preisnitz, 1972. Trans. Cave Research Group of Great Britain, p. 154)

of E are not necessarily fixed for each solute and they will depend upon the nature of the solvent in question (as discussed in the section on organic acids, 3.4). The interplay between chemical reaction rates and water residence times is a crucial factor in influencing which chemical element will be removed in solution from the soil minerals. Dynamic factors are therefore a vital consideration and simple notions of the solubility of minerals in closed systems are not enough.

5.3.2. The rate of water flow in soils

Just as solubility is a complex topic, so the residence time of water in soils is more difficult to assess than would at first appear. The pore spaces in soil, through which water travels, vary considerably in size and connectivity. They range from the microscopic intergranular spaces to pores of 0.5 to 2 mm in diameter. Moreover, fissures and subsoil pipes of 2 to 3 cm width and even larger may be present. Taking this sequence of increasing size, water will be able to travel through these pore sizes with increasing ease. However, more important factors are not simply size, but connectivity and tortuosity. In reality, the amount by which pores are interconnected and the tortuosity of the channels which arise will act to constrain or encourage water flow. On the upper end of the scale rapid transport will mean that the water will have a low residence time and a correspondingly limited opportunity to react chemically with the material through which it is passing. At the lower end of the scale the water in the smallest pores will be stationary for periods long enough for chemical equilibration to occur. The water may, in fact, because of high surface tension, be stationary for so long that it may never move at all. In this

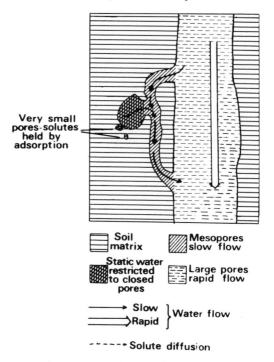

Very small
pores-solutes
held by
adsorption

▤ Soil matrix		▨ Mesopores slow flow
▦ Static water restricted to closed pores		☰ Large pores rapid flow

⟶ Slow ⎫
⟹ Rapid ⎬ Water flow
            ⎭

- - - - - • Solute diffusion

FIG. 5.11.   Diagrammatic representation of various sizes of soil pores with static water, slow flow, and rapid flow

case chemical diffusion through static water may be an important process (see Section 3.3.3). This is illustrated in Fig. 5.11. The role of pore size as an influence upon the rate of water movement in soils, often with reference to chemical equilibriation, is discussed by Anderson and Bouma (1973), Beven and Germann (1982), Bouma and Anderson (1973), Bouma *et al.* (1979), Bouma *et al.* (1981), Burt and Trudgill (1985), Germann and Beven (1981), Thomas and Philips (1979), and Trudgill *et al.* (1983).

As well as pore size and connectivity the nature of the water input will also be important, as mentioned above. Water flow rates, and residence times, will be related to the frequency and intensity of rainfall events (see Section 5.4). There are two alternative processes: one is alternate storage and displacement, the other continual flow. Storage in the soil will occur between rainfall events and displacement, or shunting of the water will occur during rainfall events. Much of the water output at slope foot drainage sites is water which has been shunted out by incoming water. It is only where rainfall is continued that continuous flow systems develop. Slope foot drainage systems can be divided into two: base flow, which is a continuous drainage of small amounts, and peak flow, the occurrence of which is directly related to the occurrence of input water (for a fuller discussion see Weyman, 1975).

Dissolution–water residence systems in soils may therefore be a matter of (1) chemical diffusion through static water in micropores, (2) storage in temporarily closed systems and displacement by successive rainfall events, (3) slow flow (either unsaturated or saturated) in medium-size pores during moderate rainfall events, and (4) rapid flow in the larger pores under conditions of sustained high rainfall. The interpretation of patterns of solute variation in slope foot drainage waters is thus complex, but certainly there will be two sets of important factors influencing the patterns—rainfall type (frequency and intensity) and soil pore size distribution and connectivity.

5.3.3. *The route of drainage waters*

A final set of complicating factors is that concerned with infiltration characteristics of the soil surfaces and the subsurface horizons and the consequent route which water takes over or through the soil. These topics are discussed in Calver *et al.* (1972), Carson and Kirkby (1972), Kirkby (1969), Rose (1962), and Weyman (1970, 1975). If the amount of water input to a soil layer is greater than the infiltration capacity of the layer then water will tend to pond up at that layer, or flow laterally (or both, in succession). Infiltration excess overland flow will thus occur if this situation arises at the surface (Pearce, 1976) and lateral subsurface flow will occur if this arises lower down in the soil profile (Fig. 5.12).

If infiltration excess overland flow occurs, the opportunity for contact with soil material will obviously be limited. This is one reason why slope foot

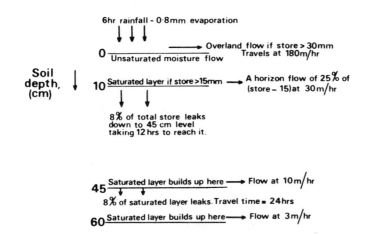

FIG. 5.12. Model of vertical soil profile under rainfall input. Diagram shows how vertical infiltration rate influences lateral water flow, saturation at discontinuities leading to lateral flow at various levels and at various speeds (from Calver *et al.* 1972)

storm drainage waters may be diluted. It would not be a matter of rapid flow through the soil, and thus little time having been permitted for chemical equilibriation (as shown in Fig. 5.3), rather it would be a matter of limited opportunity for contact with dissolvable materials. If, however, the soil 'A' horizon has a high permeability the water will tend to flow through this into the soil, picking up organic acids and chelates as it does so. Then the lower layers of the mineral soil may be contacted as waters percolate downwards. If the lower, mineral layers of the soil are less permeable than the 'A' horizon, the water will tend to flow laterally above the mineral soil, yielding slope foot drainage waters of high organic matter content but low mineral solute content. Alternatively, the mineral soil horizon may be permeable and in this case the water will percolate downwards, giving an opportunity for mineral contact and the picking-up of solutes. The water may then flow into a permeable bedrock, or, if the subsoil 'C' horizon of the soil is relatively impermeable, the water will flow laterally downslope, yielding slope foot drainage waters of a high mineral solute content. Thus the route which water takes over or through a soil mass is of vital importance in controlling solute removal from slopes (Fig. 5.13). Finally, it may be added that topography, as well as infiltration, can be an important influence upon the route which water takes on a slope. Hollows will tend to encourage overland flow while spurs and hummocks will tend to encourage infiltration (Fig. 5.14).

5.3.4. Summary

In summary, there are three key variables in the solute removal system:

1. Opportunity for mineral–water contact.
2. The length of time of contact.
3. Chemical reaction rates.

These, in turn, are controlled by:

1. The infiltration capacity of each soil horizon.
2. Rainfall type (intensity and frequency).
3. Soil pore size distribution and the connectivity and tortuosity of channels.
4. Topography.
5. The solubility of the mineral in the solvents available.
6. The transport of reactants to, and products from, the reaction sites.

5.4. The influences of rainfall regimes

Differences in rainfall frequency and intensity will be important in understanding differences in the removal of solutes. Although rainfall naturally

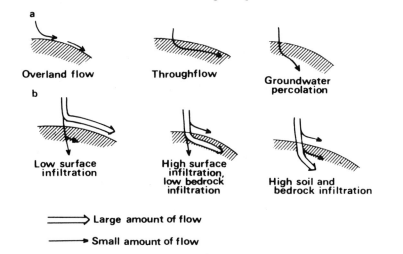

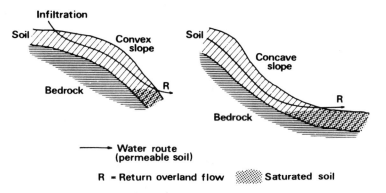

FIG. 5.13. Overland flow, throughflow, and groundwater flow (a) Routes (b) Proportions with differing infiltration rates

FIG. 5.14. The relationship between topography and water route

occurs in events of varying frequency and intensity it will be useful to model four extreme cases (Fig. 5.15). Rainfall type 1 has low intensity and low frequency. Assuming all other factors to be constant, this type of rainfall will be solutionally inoperative. Each rainfall event may instigate the solution of solids and the water will be resident in the soil long enough for most of the mineral constituents to dissolve. However, with this type of regime, the small amounts of water supplied will probably mean that removal will not occur. Thus it is probable that an oscillating system of dissolution and precipitation may be present.

Rainfall type 2 also has a low intensity but the frequency is higher. Here water will be supplied constantly enough for flow to occur and for solutes to be removed. A constant supply of light but steady rainfall will mean that soil

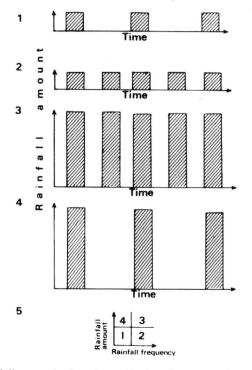

FIG. 5.15. Rainfall types: 1. Low intensity, low frequency. 2. Low intensity, high frequency. 3. High intensity, high frequency. 4. High intensity, low frequency. 5. General relationship

water flow is slow enough for even the slowly dissolving mineral constituents to dissolve but fast enough for solute removal to occur. Congruent solution of most mineral constituents can be expected in this situation.

Rainfall type 3 represents the optimum for the removal of the rapidly dissolving constituents but the opposite for the removal of slowly dissolving mineral constituents. Flow will be rapid enough for removal to be facilitated but it will, in fact, be so rapid that only those minerals possessing quick dissolution reactions will be removed to any great extent.

Rainfall type 4 will encourage the dissolution of all constituents as the residence time of the water between events will be long enough for congruent solution to occur. Type 4 may be more efficient than type 2 as greater flushing may occur.

In summary, assuming free drainage and free infiltration, as intensity and frequency rise, the solution of rapidly dissolving minerals will be favoured. As they fall, congruent solution will be favoured (unless rainfall drops below a threshold where removal is so sluggish that reprecipitation occurs).

Superimposed upon these theoretical patterns will be the factors of drainage and infiltration. Drainge discontinuities within the soil and drainage deficiencies, for example downslope, will limit the effect of rainfall events shown in type 3. Up to a certain extent the deficiencies will increase the soil water residence time (and therefore solute concentration) but over this point overland flow will occur. Return overland flow will obviously contain mineral ions in solution but infiltration excess overland flow will have lost its opportunities for solutional contact with the mineral constituents of the soil. It is now pertinent to model the solutional effects of the different types of rainfall under contrasting conditions of drainage and infiltration.

Simply, under high infiltration, the pattern described above will still hold. However, under low infiltration conditions type 1 and 4 will be more effective than types 2 and 3 respectively because the lower frequency of the rainfall allows for greater infiltration into the soil from surface storage. Of all the rainfall types type 1 now becomes the most effective in the removal of mineral solutes as this is least liable to saturate the surface and lead to overland flow. The other types will be prone to flow overland and thus will have minimal contact with the mineral soil.

Similarly, under good drainage conditions the initial pattern described will hold but, under conditions of poor drainage from the mineral soil and subsoil, type 4 will be more effective than type 3. Type 2 will be more effective than either 4 or 3. Type 1 will be the most effective as the spacing of the rainfall events will allow for slow percolation losses between events. If drainage is very poor then the rainfall types become largely irrelevant.

It is now possible to ascribe limitations to the dissolution/leaching model. Under conditions of good drainage and good infiltration the model will simply be rainfall limited. Otherwise dissolution and removal will be infiltration limited, drainage limited, or both.

Water storage in soil organic horizons is a further factor to consider. If infiltration is low then this becomes less relevant but if infiltration tends to be better then storage in organic horizons may become important. If water storage capacity in the organic horizons is high then a more frequent type of rainfall regime will be the most efficient at solutional removal since pick-up of organic weathering potential will be high and water available for leaching will be high; under low storage the converse will be true. The importance of storage will naturally be reduced if subsoil drainage is poor. It will become very important if layers of organic matter overlie relatively porous material.

Reference may usefully be made to some of the published accounts of fieldwork, theory, and models of both soil water movement and leaching in soil profiles. The movement of chemicals in soil water is discussed by Boast (1973) and Kurtz and Melstead (1973). Gardiner and Brooks (1957) present a detailed mathematical description of the theory of leaching. Similar mathematical models are given in the works of Lindstrom and Boersma (1971, 1973) and Lindstrom *et al.* (1971). Computer models for transport processes in soils are given by de Wit and van Keulen (1972) and for accumulation and leaching in soils by Frissel and Reiniger (1974). The paper by Scrivner *et al.* (1973) combines daily climatic data with a knowledge of dilute solution chemistry in order to predict soil profile formation, while Massey and Jackson (1952) and Kemper *et al.* (1970) discuss selectivity and differential mobility of chemical elements in the soil profile. Field measurements of water conductivity, acidity, and rate of flow are described by Cole (1968) and, lastly, four papers deal with lysimetric measurement of soil masses together with chemical movement in soils; they are Bourgeois and Lavkulich (1972), Cole *et al.* (1961), Marshall *et al.* (1973), and Upchurch *et al.* (1973).

5.5. Interrelationships of factors

There are close interrelationships between rainfall, organic matter decomposition, leaching, and weathering (Likens *et al.*, 1970; Crompton, 1960). In some ways the relationships are somewhat paradoxical because high rainfall need not necessarily lead to a high potential for leaching. This is because of the interrelationships between soil moisture and organic matter decomposition. If rainfall input to a soil is very high, this will act to decrease soil aeration. In turn, this will be detrimental to the decomposition of organic matter and thus lead to a build-up of a thick peaty surface horizon. Such horizons often tend to have low infiltration characteristics. Thus, rather than an organic accumulation helping to increase weathering by the provision of organic reactants, the organic accumulation helps to divert rainfall to an overland flow route, well away from the subsurface mineral soil.

A further factor is that if, initially, leaching is operative, the resultant tendency towards a low nutrient status will discourage and limit the activities of the soil biota. Thus, as well as poor aeration, this feedback mechanism will be another factor acting to discourage the decomposition of soil organic matter by soil biota.

It can thus be suggested why, in high rainfall, upland regions of Britain, fully leached podzols do not often occur (Curtis *et al.*, 1976, endpapers). Leaching by chelates is not a dominant process because (*a*) decomposition is not advanced enough to produce a large supply of organic derivatives and (*b*) much of the water is retained in peaty surface layers or diverted overland. Peaty-gleyed podzols therefore often occur in high rainfall regions. Characteristic podzols only occur in well-drained situations, often in the drier areas of the British Isles, for example, in the New Forest area in southern England on sandy or story parent materials or, locally, on sands in areas such as the East Anglian Brecklands or on the North Yorkshire moors. Podzolization in England is reviewed by Dimbleby (1962) and the review is based partly on earlier works in the North Yorkshire moors (Dimbleby, 1952) and in the New Forest (Dimbleby and Gill, 1955).

In summary, leaching removal from soil can be low for two groups of reasons. The first are hydrological. Water flow through a soil may be rapid and weathering reactions may not be able to keep pace with the water flow (whatever the weathering potential). Alternatively, water may simply flow overland, away from weatherable minerals. The second group of reasons is that, irrespective of hydrological conditions, there may simply be very little to weather in the mineral material anyway (for example, quartz sand). Alternatively, weatherable minerals may be present, but factors like a rigorous climate may limit biological weathering activity. Studies of water quality in upland areas tend to suggest that soil drainage waters are generally dilute with respect to mineral solutes. Here, the factors mentioned above combine to limit leaching loss. In particular, the formation of a peaty surface and the presence of quickflow pipes play important roles. In these situations leaching will be encouraged if surface organic layers are removed and rain-water contact with the mineral matter is facilitated.

Leaching will be encouraged in areas of moderate rainfall, where organic matter decay is at an optimum for infiltration and water retention and for the maximum liberation of weathering potential. These conditions will be fulfilled in areas of moderate drainage, suitable moderate temperatures, moderate to high nutrient status, and medium soil texture and permeability as discussed above. The question then arises of why soils in these situations do not necessarily become rapidly podzolized. The answer lies in the action of mechanisms which return nutrients from the lower soil horizons towards the surface horizons. It is probable that, in these suitable conditions, soil fauna and vegetation recycling will act to return nutrients to surface layers to offset leaching. The corollary of this is that if the vegetation is removed then leaching will advance rapidly and, under a system of agricultural management, fertilizer application will become necessary.

At this stage in the discussion it will clearly be relevant to broaden the scope of the modelling to include the topic of nutrient cycling by soil fauna and vegetation. This, then, is the topic of the next chapter.

Further reading

Equilibrium chemistry, weathering and leaching, infiltration and soil throughflow, ecosystem nutrients, and hydrology. (Burt and Trudgill, 1985; Carson and Kirkby, 1972, Ch. 9; Crompton, 1960; Gould 1971; Likens *et al.*, 1970; Trudgill, 1986; and Weyman, 1975.)

6 Nutrient Cycling

But sir, what experiments could *dis*prove your hypothesis?

JOHN R. PLATT

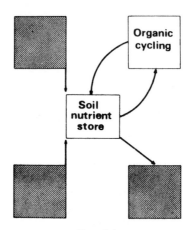

FIG. 6.1.

6.1. Introduction

It is difficult to discuss nutrient cycling *per se* without considering cycling in relation to inputs and outputs in whole systems; therefore, since whole systems form the topic of the next chapter, this chapter will focus on the retention mechanisms involved in cycling. The retention of nutrients by soil and vegetation systems, in the face of leaching losses, can take two forms. One is the redistribution of leached constituents from the lower soil profile to the upper by the activities of the soil biota. The other is the cycling of nutrients by higher plants. In the former the activities of burrowing animals are of considerable importance in physically moving soil, containing nutrients, from one part of the soil profile to another. Usually, this operates in the opposite direction to leaching losses. In the latter, losses of nutrients from the soil nutrient store by plant root uptake are offset by the return of nutrients from decomposing leaves, stems, and branches. In mature eco-systems, such return is liable to be close to the amount of uptake, especially as collapse and decay of mature tree trunks will be involved. In younger, disturbed ecosystems, then such a return of mature trunks will be lacking and there is liable to be an imbalance, with uptake and biomass storage exceeding return. In addition to biomass decay, foliar and stem leaching by rainfall and throughfall can also act to return nutrients to the forest floor. The fate of the returned nutrients depends upon the efficiency of soil retention and root uptake mechanisms. Where there are few exchange sites on soil clay-humus complexes, or where exchange sites are occupied by H^+ or Al, or where root uptake is limited, then leaching losses are liable to dominate over retention and recirculation in the soil and vegetation system. The activities of soil organisms are considered below; the major processes of nutrient cycling by vegetation are outlined in Section 6.3. and nutrient sources and budgets in Section 6.4.

6.2. Soil mixing by organisms

Soil biological activity, especially the actions of earthworms and other burrowing animals, such as moles, can be very effective in moving subsoil to the surface. As its most marked, this upward movement of soil material can effectively negate the downward leaching of nutrients and acts to keep the topsoil nutrient rich. The activities of animals in soils are reviewed by Abaturov (1972), Thorp (1967), Edwards *et al.* (1970), and Russell (1973, Ch. 11).

It is of interest to note that biological activity in a soil is influenced by soil acidity and nutrient status. Populations and activities of soil biota are higher in nutrient-rich neutral and alkaline soils than in nutrient-poor acid soils. It can therefore be suggested that soil organisms may operate a negative feedback mechanism, keeping the nutrient-rich soils rich in nutrients by their activities. Neutral and alkaline soils may therefore remain so partly as a result of animal activity bringing mineral-rich material to the surface from lower horizons. However, below a minimum level, soil nutrient status and acidity may become low enough to begin to inhibit biological activity. As animals, such as earthworms and moles, begin to decrease in activity and number a positive feedback may occur. That is, that a slightly acid soil may discourage biological activity which, in turn, will lead to a diminished amount of upward soil movement. This will further reinforce the acidity of the soil and this will further compound the decrease in biological activity. The trend to acidity and leaching loss of nutrients thus progressively gains the upper hand in the system.

In a study of limestone and related soils, Bullock (1971) suggests that this kind of mechanism may operate in the transition between lime-rich and lime-poor soils. The transition is often marked by a sharp contrast between limestone soils (rendzinas and calcareous brown earths) and acid soils (podzols and acid brown earths). It can be suggested that this contrast is heightened by the feedback mechanisms discussed above, the lime-rich soils being reinforced in their alkalinity by the activities of soil animals while progressive leaching and loss of soil biota operate in the acid soils. In rendzina soils of pH in the range of 6 to 7 Bullock reports a high population of earthworms, of up to 100 per square metre. The species present were dominantly *Lumbricus festivus* and *Allolobophora chlorotica*. These thoroughly mix mineral and organic matter throughout the whole soil. In wetter and more acid soils of pH in the region of 4 to 5 the earthworm populations are either not present or of small significance.

Animal activity varies in its scope and significance. Earthworm activity is usually of significance in soils of circumneutral and alkaline soils (Jackson and Raw, 1966, p. 19; Wallwork, 1970), and their activities have been demonstrated by their artificial removal using pesticides. Organic matter rapidly collects on the surface and the soil becomes poorly structured (see, for example, Curtis, *et al.*, 1976, p. 284 and also Vimmerstedt and Finney, 1973). Locally, other organisms may become important and in semi-arid or tropical soils animals such as ants, termites, or burrowing animals may be of importance.

6.3. Vegetation processes and decomposition

The key processes to consider are the ways in which nutrients are taken up into vegetation and then returned to the soil. In the former, the distribution of root networks in the soil is important and, in the latter, the important processes are leaching from the above ground parts of the plant by rainfall, grazing by herbivores, and return to the soil in faeces and the fall of dead parts of plants, including leaf litter, stems, and trunk trunks. Nutrient storage in biomass is also an important intermediate consideration. During the nutrient cycling process, nutrients are often stored in standing biomass for longer periods of time than when they are involved in decomposition and release processes. During the latter they are vulnerable to leaching and loss from the system, a factor discussed further in Chapter 7 when whole systems are considered.

Taking nutrients stored in the vegetation biomass as the starting point, whether grazing animals are involved or the fall of dead biomass, the decomposing organisms play a key role in the breakdown of dung, leaf litter, and woody tissue and the release of stored nutrients. The breakdown of woody tissue is mainly initiated by fungi and, for other organic material, a succession of invertebrate animals and soil microflora is involved.

Initially, ingestion by earthworms is especially important in leaf litter breakdown and earthworms egest finely mixed mineral–organic matter, composed of ingested soil particles and digested organic matter. The earthworms casts may be deposited above or below ground and they are generally far more nutrient rich than the surrounding soil, the principle source of the

Table 6.1   *Nutrient content of earthworm casts compared with surrounding topsoil*

Material	Available nutrients (ppm dry soil)				
	Ca	Mg	K	P	NO3−N
Topsoil	1990	162	32	9	4.7
Wormcasts	2790	492	358	67	21.9

Source: Russell (1973)

nutrients being the ingested organic matter. Table 6.1. shows some data comparing the nutrient content of earthworm casts with the surrounding soil. In the casts the nitrate–nitrogen content is 14 times higher than the soil, potassium 11 times higher, phosphorus 7, magnesium 3, and calcium 1.4 times higher.

Other invertebrates, especially springtails (*Collembola*), mites, beetles, and various insect larvae, ingest and comminute leaf litter. Smaller organisms, especially bacteria and also some fungi, are responsible for the final breakdown of comminuted leaf litter, earthworm casts, and animal dung.

The amount of nutrients found in stems, branches, and leaves varies from species to species and, within one species, with the age of the plant. The major effects of nutrient cycling may therefore vary considerably according to the nature of the vegetation present, for instance, Nihlgard (1972) documents differences between spruce and beech vegetation, the contents of calcium and magnesium, for example, being lower in the spruce litter. The release of nutrients from decomposing branch and leaf material has been studied by Gosz *et al*. (1973). In this work it was noted that the maximum decomposition rate for branch and leaf tissue occurred during the summer months. Nutrients were released by two mechanisms, leaching from the tissue and decomposition of the tissue. Potassium and magnesium were rapidly leached from the tissue while calcium was released by decomposition. This was clear since, in the former case, element losses were in advance of structural weight loss but, with the latter, element losses correlated closely with dry weight loss. The quantities of nutrients released from leaf and branch litter were very large when compared with the small total output from the system. This is taken as evidence that there is a very strong biological retention of nutrients by cycling processes and that by uptake of released nutrients they are conserved within the ecosystem.

Leaf type is an important factor in decomposition and also in foliar leaching. Leaves with a thick cuticle (outer covering) and a waxy surface (such as evergreens), tend to be more resistant to foliar leaching. This is also true of trees with thicker bark which inhibits the diffusion of nutrients out from the phloem and the subsequent losses in stemflow. The thicker leaves also tend to be less susceptible to insect attack and take longer to decay once fallen. Small (1972) also suggests that the nutrients in the thicker, more long-lived ever-green leaves may be retranslocated back into the twig for further, later use before leaf fall, especially phosphorus, nitrogen, and potassium. Thus, these types of leaves represent a system for lower, or at least slower nutrient release back to the soil, with greater nutrient conservation in the biomass. Softer, more palatable, more easily decomposible leaves release their nutrients more quickly and are also more susceptible to insect attack, which is an important mechanism for nutrient release, especially under heavy infestations (Swank, 1986).

Once released by decomposition, the most important factor in nutrient recycling is the uptake by plant root systems. Root systems may be wide-spread and deep, scavenging nutrients from a wide area, or they may be concentrated in a root mat, especially under and in a surface layer of humus and decomposing litter. Here, the rapid growth of a network of fine, feeder roots is important in the uptake of nutrients and their retention by the nutrient cycling system.

6.4. Nutrient budgets

Some cycling systems appear to be relatively closed, with efficient retention within the soils and vegetation system, while others are relatively 'leaky' with considerable losses to runoff waters—which can then only be replaced by bedrock weathering, atmospheric inputs or fixation. Such evaluations can be made by the study of amounts of nutrients present in different components of the system or, most effectively, by isotopic labelling of particular nutrients.

In one such study, Stone and Kszystyniak (1977) found that after 9 years, 60 per cent of labelled potassium applied to a stand of *Pinus resinosa* trees on sandy soils in northern New York State was still cycling within the ecosystem and after 23 years, 40 per cent was still cycling, suggesting that losses are of the order of 40 and 60 per cent respectively. Working in the Hubbard Brook northern hardwood forest in the USA, Likens *et al.* (1977) suggest that the quantities of calcium circulating within the ecosystem are much greater than those lost by leaching, as shown in Table 6.2. Root uptake is 62.2 kg ha^{-1} a^{-1}, some of which is stored within the biomass and the return to the forest floor in litter fall, throughfall, and stemflow is 47.4; these amounts in circula-tion are considerably greater than the losses to streams at 13.9 kg ha^{-1} a^{-1} showing that cycling is an important mechanism of conserving nutrients in the soil and vegetation system. This is despite an overall export, with hydro-logic export minus precipitation input of 2.2 being 11.7 kg ha^{-1} a^{-1}. Much of this must be replaced by weathering, at 21.1 kg ha^{-1} a^{-1}, part of which will also be taken up by the vegetation and act to offset biomass storage.

In other situations, Jordan and Herrera (1981) present data on calcium for

Table 6.2 *Principal components of the calcium cycle, northern forest ecosystem, USA*

	Fluxes (kg ha^{-1} a^{-1})	Accretions (kg ha^{-1} a^{-1})	Nutrient Stores (kg ha^{-1})
Precipitation	2.2		
Root uptake	62.2		
Root biomass		2.7	101
Above ground biomass		5.4	383
Throughfall and stemflow	6.7		
Litter fall	40.7		
Forest floor		1.4	370
Forest floor to soil	42.2		
Weathering input to soil	21.1		
Available in soil			510
Hydrologic export	13.9		

Source: Likens *et al.* (1977)

Table 6.3 *Principal components of the calcium cycle in tropical and temperate ecosystems. Data in kg ha^{-1} a^{-1}*

		Temperate Eutrophic	Tropical Eutrophic	Temperate Oligotrophic	Tropical Oligotrophic
Precipitation	(P)	10.5	21.8	3.3	16.0
Runoff	(R)	27.4	43.1	9.7	2.8
P − R		−16.9	−21.3	−6.4	+13.2

Source: Jordan and Herrera (1981)

a range of ecosystems which suggest that losses (precipitation input minus runoff output) range from 6 to some 20 per cent (Table 6.3.) while for an Amazonian tropical rain forest site, there was a net gain of some 13 per cent indicating a very tight conservation mechanism in the latter and the retention of atmospheric inputs of nutrients. Duvigneaud and Denaeyer-de Smet (1970) present data for a closed circulation system where uptake from the soil is balanced by woody increments and litter fall for beech and pine (Table 6.4) while Smith (1984) lists losses and gains for calcium and potassium in northern forested ecosystems (Table 6.5). Calcium and potassium generally show a greater leaching output than atmospheric input but again cycling is an

Table 6.4 *Nutrient cycling in beech* (Fagus sylvatica) *and pine* (Pinus sylvestris). *Data are in kg ha^{-1} a^{-1}*

	K	Ca	N	P
Beech				
Uptake from soil	14	94	50	12
Wood increment	4	13	10	2
Litter fall	10	81	40	10
Pine				
Uptake from soil	6	29	45	4
Wood increment	2	10	10	1
Litter fall	4	19	35	3

Source: Duvigneaud and Denaeyer–de Smet (1970)

Table 6.5 *Nutrient budgets for selected forest ecosystems. Data in kg ha^{-1}a^{-1}. Data in brackets are for plant uptake (where available)*

Ecosystem	Calcium			Potassium		
	Input	Output	Net	Input	Output	Net
Beech (Germany)	12.4	14.1	−1.7	2.0	1.6	+0.4
Douglas Fir (USA)	2.3	50.3 (45)	−48.0	0.11	2.25 (23)	−2.14
Deciduous (Hubbard Brook, USA)	2.2	13.7 (53)	−11.5	0.9	1.9 (53)	−1.0
Deciduous (Coweeta, USA)	4.8	7.7 (75)	−2.9	2.1	5.6 (62)	−3.5

Source: Smith (1984)

important retention mechanism as nutrient uptake levels by the vegetation are often greater than the losses.

Clearly, part of the differences in the results obtained by different workers depend upon disturbances: more stable systems are liable to have a balance between uptake and return or greater output than input, while disturbed systems are liable to be aggrading, with increased storage in standing biomass from year to year, and greater inputs than output. Disturbances can take the form of the removal of timber, with its stored nutrients. This can represent a considerable loss and Smith (1984) quotes ranges of from 90–769 kg ha^{-1} for calcium, 52 to 219 for potassium, 4.8 to 17.5 for phosphorus, and 84 to 386 for nitrogen, the lower figures being for pine and the higher for oak. Replacement of mature trees by regrowth will obviously lead to further incremental storage in the biomass, presumably from precipitation and weathering inputs. Fire can be seen as another disturbance which acts in the same way, with loss of nutrients in smoke. In addition, losses of nutrients by leaching after a fire can also be important as they can be easily solubilized from ash. Smith (1984) quotes fire loss figures of 12.5 and 2 kg ha^{-1} for calcium and potassium respectively for a 12 year old *Calluna* (Heather) heathland. This is considerable when compared to the nutrient content of the vegetation (33 Ca and 4 K, kg ha^{-1}) and litter (15 Ca and 4 K, kg ha^{-1}). It is clear that in many cases, there are large amounts of nutrients stored in and flowing through the biomass, disturbance of which represents a considerable loss to the soil and vegetation system. This may be especially true on nutrient poor soils.

Johnson and Herrera (1981) propose that oligotrophic systems, that is those vegetation stands on nutrient poor soils, are far more conservative of nutrients than those on eutrophic (nutrient rich) soils. This especially applies to tropical rain forests on nutrient poor soils where evergreen, thick cuticled waxy leaf types (scleromorphs) and dense root mats combine to minimize nutrient losses and to maximize retention and cycling. Here also, other nutrient-conserving mechanisms were evident. Thus, leaves with drip tips exhibited more rapid rainfall runoff, thus minimizing opportunities for foliar leaching. Many leaves also contained toxins which were deterrent to herbivores, thus minimizing grazing losses. They also suggest that mychorrhizal fungi (those associated with plant roots) were important in the transfer of nutrients from the decaying leaf to the root. They also quote the work of Stark and Jordan (1978) who performed experiments with labelled isotopes. These authors showed that direct adsorption was a major process: 99.9 per cent of labelled ^{45}Ca and ^{32}P sprinkled onto root mats was immediately adsorbed, with only 0.1 per cent leached through to runoff water. When the isotopes were applied in the form of labelled leaves, no leaching was observed.

6.5. Generalizations

There is considerable variation in the breakdown and uptake processes, depending on the vegetation type and the degree of disturbance to the system. Detailed descriptions of nutrient cycling are reserved until Chapter 7 when vegetation and elements have been specified (see, for example, the sections on calcium cycling, 7.5; and on silica cycling, 7.6). At this stage of the discussion, only a few broad generalizations will be possible. Additional reading may be obtained from Ovington (1962) and Reichle (1970). Other useful sources for reference include Deevey (1970), Bormann and Likens (1967, 1969, 1970), and Chapman (1967).

Plant litter decomposition, cycling, and the elemental content of plants are further discussed by Blow (1955), Bocock *et al.* (1960), Bornkamm and Bennert (1970), Dickinson and Pugh (1974), Duvigneaud and Denaeyer-de Smet (1970), Gilbert and Bocock (1960), Tolgyesi *et al.* (1968), and Williams (1953).

Generalizations can be made on two topics: the amounts of nutrients involved and the types of nutrients involved. In terms of simple amounts, root uptake and the return of nutrients in litter are closely related to plant productivity (Woodwell and Whittaker, 1967; Rodin and Bazilevich, 1967). Given any one particular plant species, growth conditions (especially climate) are of primary importance in controlling productivity. Cycling can, then, be modelled in gross terms by the use of environmental factors which predict productivity. It is also of interest to note that nutrient supply itself is one factor which contributes to the control of plant productivity. Again, an element of mutual interaction with the system is an important factor.

It is difficult to go beyond these generalizations without consideration of specific context. Accordingly, it will be most useful to proceed to the next chapter and to discuss the system as a whole. In summary, nutrient cycling acts as a retentive mechanism, which acts to offset leaching losses. It appears that nutrient release from litter decomposition may be a supply for plant root uptake. This represents a substantial measure of nutrient conservation in soil and vegetation systems.

Further reading

Nutrient cycling; woodland ecosystem, production, and mineral cycling. (Bormann and Likens, 1967, 1970; Likens *et al.* 1977; Ovington, 1962; Rodin and Bazilevich, 1967; Smith, 1984; Swank, 1986.)

III
Nutrient Systems: Models of Whole Systems

7 Models of Input, Output, and Cycling in Whole Systems

Nature speaks to us in detail and only through detail can we find her grand design.

BRONOWSKI, talking of Leonardo da Vinci

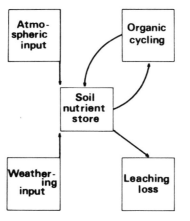

FIG. 7.1.

7.1. Introduction

Modelling of whole systems is a difficult topic because there are few studies which have taken an overall viewpoint and yet stressed each component in equal detail. A useful comprehensive review of forest nutrient cycling is provided by Waring and Schlesinger (1985) and nutrient cycling in tropical forest ecosystems is reviewed by Jordan (1985). One of the more well-known overall studies is that by Likens *et al.* (1977) on the biogeochemistry of a forested ecosystem at Hubbard Brook in the northeastern USA. Other studies have tended to stress various aspects, for example the ways in which biological processes control hydrological output of solutes (Swank, 1986). Given the current state of knowledge, it is possible to undertake two things, one to discuss theoretical models of whole systems and the other to assess the available data which will help to quantify and assess these models.

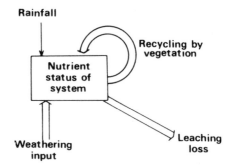

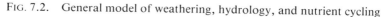

FIG. 7.2.   General model of weathering, hydrology, and nutrient cycling

7.2. Models and data

7.2.1. Models

In its simplest form, the basic model is the one which has been used at the Chapter heading (Fig. 7.1.) and which can now be reconsidered in Fig. 7.2. This shows how four factors influence the soil nutrient store, with nutrient removal in drainage waters (Chapter 5) acting in opposition to the inputs from weathering (Chapter 3) and rainfall (Chapter 4), and the net retention from nutrient cycling (Chapter 6).

In more detail, it can be seen that the input of nutrients from the atmosphere is an independent variable, but weathering input, recycling by vegetation, and leaching losses may all be somewhat related to each other and to atmospheric inputs. Thus, for example, weathering might be encouraged by plant nutrient uptake and leaching losses but decreased by cation inputs in rainfall, and nutrient cycling can act to decrease losses by leaching. There thus exists an element of feedback in the system and this can be summarized in Fig. 7.3. It is suggested that leaching acts to decrease soil nutrient status, shown in the centre of the diagram; such a decrease would then trigger a greater rate of weathering, acting to replenish the soil nutrient status. A high soil nutrient status could increase the amount of biological cycling, the actual amount depending upon the way in which climate and vegetation combine to give high or low biological productivity. Increased cycling associated with high biological productivity would then increase the amount of nutrients in the soil store.

A key factor is whether or not there is a store of weatherable minerals in the soil and bedrock which could supply the available nutrient store in the soil. If not, losses in leaching cannot be resupplied by weathering and the amount of cycling will be the most significant factor in the resupply of available soil nutrients. If weatherable minerals are present, plant uptake and leaching can be offset by weathering and cycling, although obviously present, will not be so significant in replenishment of soil nutrients.

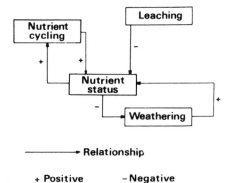

FIG. 7.3. Simple model of the feedback relationships between the nutrient system factors.

In reality, this replenishment of the soil store by weathering would involve the movement from the minerals to the clays and soil solution, as suggested in Fig. 2.3., and as shown in Fig. 7.4a. Here, the processes and flows of cations between the major stores, inputs, and outputs are shown. These processes would themselves be subject to a number of feedback relationships and external controls. A more detailed diagram of these relationships is suggested in Fig. 7.4b. In this figure, an arrow has the meaning of 'has an influence upon', rather than indicating a flow of materials as in Fig. 7.4a.

If an increase in the amount, or action, of one factor leads to an increase in the amount, or action of another then the first factor in one box is linked to the second by a positive arrow. If, however, the relationship is reciprocal and an increase in the first factor leads to a decrease in the second, then a negative arrow has been used. It can be seen that there are only two negative arrows—that for the effect leaching has upon the nutrient status of the soil solution and that for the effect nutrient status has upon weathering potential. Negative feedback can occur in the system for, as the nutrient status of the system decreases, the potential for weathering will increase and, assuming that weatherable minerals are present in the rock and soil, weathering will occur to offset the lowering of nutrient status. In this way nutrient status may be maintained.

Biological cycling is especially important in reinforcing nutrient status. The role of biological cycling in nutrient retention in ecosystems has been discussed by Bormann *et al.* (1969), Damman (1971), Edwards *et al.* (1970), Fortescue and Marten (1970), Horrill and Woodwell (1973), Likens *et al.* (1970b), Redfield (1958), and Siccama *et al.* (1970). Much of the American work cited here has involved the use of herbicides, clear-felling, or radiation to remove the vegetation component. With the loss of natural cycling processes, marked increases in nutrient losses as stream solutes have been noted.

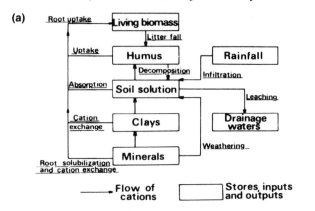

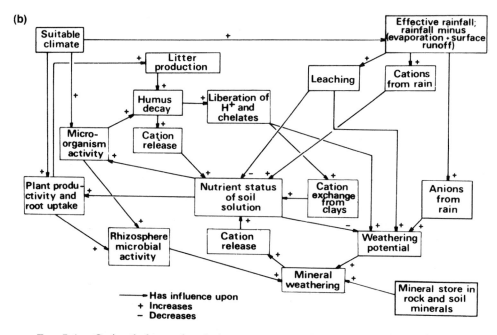

FIG. 7.4. Cation balances in whole systems (a) main components and processes, arrows indicating fluxes; (b) feedback web relationships, arrows indicating relationships which increase (+) or decrease (−) receiving component.

Other important feedback relationships which emerge from this model are that leaching will act to increase the weathering potential, encouraging the release of cations by weathering. However, the actual store of minerals present in the rocks and minerals in the soil-rock system could become a limiting factor upon cation release. Clearly, if the mineral store is low, then

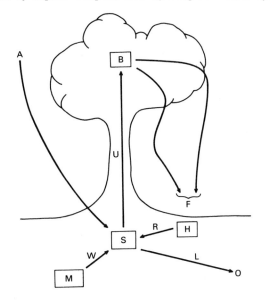

FIG. 7.5. Simplified diagram of the main fluxes (arrows) and stores (boxes) in a forested ecosystem.

The fluxes of nutrients are indicated by arrows, and stores are indicated by boxes, both with shorthand notations as follows:

Fluxes:

A = Atmospheric inputs—wet and dry deposition.
W = Weathering inputs from bedrock and soil minerals.
U = Uptake by plant roots.
D = Dead biomass fall, litter, stems, and tree trunks.
T = Throughfall and stemflow.
F = Forest floor inputs (D + T).
R = Return of nutrients from vegetation via humus to soil store.
L = Leaching through the soil.
O = Output to runoff waters.

   Stores:

M = Minerals in bedrock and soil.
B = Biomass storage—roots, stems, and leaves.
H = Humus store, nutrients in dead biomass not yet released by decomposition.
S = Soil store of available nutrients—soil solution and absorbed on clays and humus.

once it has been depleted, no matter how high the weathering potential may become, the nutrient status of the soil solution will become lower. In addition, the vegetation may well become altered to a type adapted to exist in low nutrient conditions and which may well be less efficient at nutrient cycling.

In environmental terms the primary limiting factors of this model can be identified as climate, for its influence upon effective rainfall and biological productivity, and the mineral store present in rocks and minerals. In addition, for each individual cation, its mobility in the system will differ and vary in its significance for plant growth. The biological involvement of individual elements is discussed by Burd (1925), Lee and Hoadley (1967), Nilsson (1972), Rennie (1955), and Swank (1986).

7.2.2. Data

In practice, it is difficult to provide data for such a complex diagram as Fig. 7.4b. as perhaps only two or three relationships could be studied at this level of detail. However, as discussed in the Introduction to this book, it is possible to be more holistic, looking at the whole system in general terms and at the ways in which the detail fits into the overall picture. Thus we can relate and bring together the relevant information at a slightly less complex level of resolution. A diagram, suggesting the components that can be, and have been realistically quantified is shown in Fig. 7.5.

This is a simplified diagram and minor components not specified include the decomposition of roots directly in the soil, and which can be included in R, and the immobilization within the soil profile, that is, nutrients present in L but not in O. Also, many pathways are more complex than indicated and, in practice, difficult to separate. For example A may impact on the biomass or the forest floor and thus appear in S via T and/or H. The diagram is thus a simplified representation rather than an indication of all pathways or processes.

The relationships involved are such that the soil store, S, is replenished by A, W, and R and decreased by U and L; the net cycling, C, is the balance of $-U + R$ from and to S. Thus:

$$S = A + W + R - U + L \tag{7.1}$$

or

$$S = A + W \pm C - L \tag{7.2}$$

C is liable to be negative as:

$$F = U - B \tag{7.3}$$

and

$$R = F - \Delta H \tag{7.4}$$

where Δ is used to signify a change in the store, in this case an accretion.

Other relationships are:

$$O = A + W \pm C \tag{7.5}$$

and, at equilibrium in a mature stand:

$$U = R \tag{7.6}$$

In respect of the relationship of O with A, W, S, and R, it is difficult to tell what the sources of the nutrients in O are in that they could have been derived from A and W via S if the nutrient cycle, C, is closed or from R via S if it is not closed, or some combination of the two (as discussed further for calcium).

Having specified the main relationships we can now take another look at the data of Likens *et al.* (1977), Jordan and Herrera (1981), Duvigneaud and Denaeyer-de Smet (1970), and Smith (1984), as shown in Tables 6.2.–6.5. The tables can provide data for many of the components shown in Fig. 7.5. and they are used below to illustrate some of possibilities and difficulties of quantifying and interpreting the system, even with purpose-built studies. Further data will also be considered in a case study of a specific environment in Section 7.3., and two specific case studies, calcium and silica, are discussed in Section 7.4.

Using Table 6.2. and further data provided by Likens *et al.* (1977), Fig. 7.6 presents a quantification of Fig. 7.5 for calcium in the northern forested ecosystem at Hubbard Brook, USA.

It is clear that the systems is not in equilibrium and that biomass accretion, ΔB, is not matched by return to the forest floor. This is what would be expected in an aggrading system. In a mature system:

$$U = R \tag{7.6}$$

or

$$U = F \tag{7.7}$$

Unambiguous data for R are not given as the authors show net mineralization both from the forest floor and the mineral soil-bound Ca + rock as one flux, 42.4 kg ha^{-1}a^{-1}. T is given as 6.7 and D as 40.7 kg ha^{-1}a^{-1}. Thus the data for Eqn. 7.7, where $F = T + D$, are not balanced:

$$U\ 62.2 \neq F\ 47.4 \qquad\qquad (\text{kg ha}^{-1}\text{a}^{-1})$$

However, if biomass storage, ΔB, is considered, a value for below ground biomass is given as 2.7 and for above ground biomass 5.4, giving a total of 8.1 kg. ha^{-1}a^{-1}. But, using Eqn. 7.3:

$$F\ 47.4 \neq U\ 62.2 - B\ 8.1 \qquad\qquad (\text{kg ha}^{-1}\text{a}^{-1})$$

which gives a shortfall of 6.7 unaccounted for. However, root decomposition

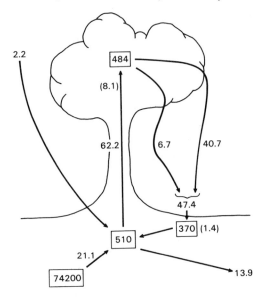

FIG. 7.6.   A quantification of Figure 7.5: calcium in fluxes and stores in a northern hardwood forest ecosystem, data from Likens *et al.* (1977). Figures in brackets represent annual accretions (Δ values). Data are in kg/ha a^{-1} for fluxes and accretions and kg/ha^{-1} for stores

is given at 3.2 and root exudates at 3.5, accounting for 6.7. It is thus clear that R = 54.1, includes root processes. Then:

$$R\ 54.1\ =\ U\ 62.2\ -\ B\ 8.1 \qquad\qquad (\text{kg ha}^{-1}\text{a}^{-1})$$

Thus U and R are balanced by considering ΔB. The last cutting of the forest was between 1909 and 1917, and the authors assume that major forest regrowth took place around 1915; with measurements in 1970, this gives an age of 55 years and it is suggested that this is insufficient to establish complete return of U in the stand as a whole by the fall of mature stems and trunks.

Some of the nutrients in F are bound on the forest floor, with F at 1.4 kg ha^{-1}a^{-1}, leaving in fact 52.7 for R. The balances then become complex as they are influenced by A inputs (2.2), weathering inputs (21.1), and hydrologic export, O (13.9) together with net mineralization from humus and soil (42.2). It is difficult to be precise about the pathways but with R at 52.7 and U at 62.2, C shows a deficit of 9.5 (8.1 ΔB, 1.4 ΔH) to be replenished from the soil store S. Inputs to the soil store other than R are A + W = 2.2 + 21.1 = 23.3 and exports other than those involved in C are O at 13.9, leaving a balance of +9.4 which will make up the deficit in C due to ΔB and ΔH, above. Thus, this also suggests that the fate of A + W is partitioned into O and U at an approximate 60 : 40 ratio. However, while the figures may

balance, the actual nutrients involved cannot be assessed because the input in
A may actually be taken up as U or equally lost as O and O can have a source
from H and so on. Only labelled isotopes could tell us which calcium ions
were which but the point of the exercise is more to show how the fluxes and
stores are related and thus how they might respond to disturbances and
management, as discussed further in Chapter 8.

Returning to other data presented in Chapter 6, as mentioned above, the
data in Table 6.3. are simply for A and O, $A - O$ telling us something about
weathering inputs but not whether they are directly from W via S, or
indirectly through C from W via U, R and S, since:

$$O = A + W \pm C \tag{7.5}$$

and only O and A are known. Three systems show a net export from the
system ($A - O$ is negative) with $W \pm C$ (i.e. $O - A$) at 16.9, 21.3, and 6.4 kg
ha^{-1}a^{-1} for eutrophic tropical and temperate sites and an oligotrophic
temperate site respectively. A tropical oligotrophic system showed a net
retention of calcium since A was greater than O ($A - O = 31.2$). Although
there is no data to partition W and C, this does suggest that increases in C
decrease losses in O.

For Table 6.4, U, ΔB, and D are given for four nutrients. Although no
other fluxes are specified, D is always less than U and ΔB makes up the deficit
$U - D$. This again suggests that the data are for aggrading ecosystems.

Lastly, in Table 6.5. again only inputs and outputs are given, and with one
exception (potassium in beech) the greater output than input suggests a net
weathering source.

The data presented are thus of interest, firstly because they show that
generalized figures like Fig. 7.5 can be quantified and secondly, they show
the relationships between the variables, suggesting the weathering source for
hydrological output losses but particularly suggesting the biological control
of nutrient losses. This appears to be especially important on nutrient-poor
sites.

The biological control of nutrient losses is dealt with in some detail by
Swank (1986). He presents data (his Table 3.3, p. 92) for $U \Delta B$, T, and D for
nitrogen, calcium, and potassium which suggest that:

$$U = \Delta B + T + D \tag{7.8}$$

of which D represents the largest component for the partitioning of U, for
example, for calcium and Coweeta Hydrologic Laboratory:

$$U\,75 = \Delta B\,23 + T\,8 + D\,44 \qquad \text{kg ha}^{-1}\text{a}^{-1}.$$

The amounts involved in A (4.8) and O (7.7) are relatively small and while
comparing A and O again suggests a net weathering input (2.9), it is clear that
most of the nutrients are involved in cycling, C, and thus disturbance of
vegetation is liable to have a marked influence on the nutrient status of the

soil, *S*, and also on the outputs, *O*. The subject of disturbance will be returned to in Chapter 8.

Given the above discussion of general relationships, the questions which can be asked are how do they vary in particular environments and with respect to specific chemical elements. Below we discuss some specific examples.

7.3. An example of a specific environment: nutrient-poor uplands

Clearly, many combinations of climate and geological substrate are possible, but for the purposes of illustration, nutrient-poor uplands have been selected. These have cool climates with high rainfall and often have steep slopes which encourage leaching. The geology includes quartzites, granites, schists, and slates and the vegetation includes heaths, grasslands, and some woodlands. Here the principle limitations upon the system are basic infertility of mineral materials and cool temperatures. Fig. 7.4b can be re-drawn for this specific environment to illustrate the chain of functioning (Fig. 7.7). These environments have a high weathering potential but leaching tends to occur because the soils are easily leached, rather than because of humus decay and liberation of organic acids. Obviously, any weathering potential cannot be effective in releasing nutrients from rocks because the rocks themselves are

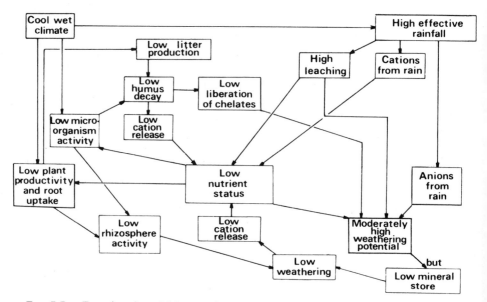

FIG. 7.7.   Functional model for nutrient system factors. Upland area, nutrient-poor rocks

poor in nutrients. Most of the nutrients, then, come from rain-water sources. As weathering does not proceed to any advanced stage clay minerals do not form readily and any weathering products that do form would be removed readily under conditions of strong leaching.

Although humus production will be low, humus decay will be severely limited, organic surface horizons will tend to form, and overland flow will be encouraged. Vegetation type will thus be of importance in affecting water route and leaching. Grassland humus, with a higher infiltration capacity, could retain more water than a greasy, peaty heath or bog humus. In actual fact, the leaching factor may be no more than a potential factor, unrealized because of the dominance of overland flow or pipe flow (see Section 5.5). Because of humus accumulation, waterlogging will be common especially in fine textured soils. Water residue times will be of little relevance as there will be few dissolvable minerals present. The water quality of drainage waters can be explained more by a consideration of pluvially derived nutrients (and their possible adsorption onto organic sites) rather than by weathering processes.

The factors of uptake (U) and return (F or R) will be minimal and will act to limit nutrients (S) and leaching (L). In terms of weathering potential atmospheric and organic contributions may be quite high but they will be little used because of the small store of weatherable minerals present. Indeed, weathering potential of drainage waters will be high as the water quality of streams from such areas will be largely derived from atmospheric and organic sources rather than from mineral sources.

7.4. Examples of specific elements

It is difficult to quantify the amounts of cations involved in plant uptake, litter return, rainfall input, drainage losses, and weathering on broad environmental levels. At the current state of knowledge it is not possible to enumerate flows and storage for particular cations for large environmental systems, much as it might be desirable. Large-scale studies of weathering products (for example, Stoddart, 1969) or of river water quality (for example, Crickmay, 1974; Davis, 1964; Douglas, 1972; Hem, 1970; Livingstone, 1963; Meybeck, 1976; McGinnis *et al.*, 1969) are available but they are not necessarily compatible with studies of rainfall input, plant uptake, and litter return. In order to achieve integration on any rigorous basis the studies involved would have to have had integration in mind from their inception so that definitions and units would have been compatible. A few such studies have been undertaken (for example, Cleaves *et al.*, 1970; Reichle, 1970) but, apart from these studies, our understanding is fragmentary.

The chief difficulty of attempts to fuse individual studies of processes and mass flow lies in the variety of purposes for which research programmes have been undertaken. In theory it ought to be possible to dovetail one study onto the next to achieve a fuller understanding of the over-all system, but differences of scale and definitions have made this problematical. For example, an ecologist may undertake a study of calcium cycling in a forest stand (Thomas, 1969), a geomorphologist a study of calcium removal from a limestone area (Smith and Newson, 1974), and an agricultural hydrologist a study of calcium loss in drainage waters from farmland (Williams, 1970). Their field areas, aims, methods, and expression of results can be so dissimilar as to make meaningful generalization impossible. In the absence of purpose-built ecosystem studies (apart from Cleaves *et al.*, 1970, Bormann and Likens, 1969, and Swank, 1986) it is only possible to do little more than point the way to integrated studies from simply discussing a variety of different works in succession, using as common themes some selected cation or defined environment. The following section attempts to undertake such a discussion and deals with selected cations. It dwells both upon existing ecosystem studies and upon attempts to bring together studies which are on a common topic, but which approach the topic from different angles.

7.4.1. Case-Study (1): The ecosystem geochemistry of calcium

Calcium is an element which is relatively easily measured in water, soil, and plant tissue by titration (Douglas, 1968; Lundegardh, 1951; Schwarzenbach and Flaschka, 1969) or atomic absorption spectrophotometry analyses of water or acid aqueous extracts of solid materials (Bisque, 1961); solid rocks can be analysed for calcium content by X-ray fluorescence (Leake *et al.*, 1969). It has been the focus of study of ecologists, hydrologists, pedologists, and geomorphologists. The purpose of this discussion is simply to ascertain how far compatible information is available upon the input, uptake, cycling, and losses of calcium, other than that from the special purpose studies already discussed earlier in the Chapter. While much of the information comes from studies of temperate lowland environments, such as deciduous woodlands, the integrated interpretation of available data is limited as the data are derived in differing ways.

One of the key issues in the discussion is the evaluation of the amounts of elements involved in cycling and uptake relative to the amounts lost by various means. The first task is thus to establish order of magnitude estimates of the various components of the system (as described in Fig. 7.5). In terms of gross solute input and output, figures for calcium in rainfall and in stream water are available (though for the latter a distinction must be made for those works which have studied limestone and those which have studied non-limestone areas).

Figures for the calcium content of rain-water commonly range from near zero to about 20 p.p.m. (Calcium as Ca^{2+}; all figures are expressed thus unless otherwise indicated), the higher figures being commoner nearer the sea. Extreme high figures can be found next to the sea, near limestone quarries and exposed limestone roadways and pathways which disperse calcareous dust into the atmosphere; figures can be as high as 50–100 p.p.m. in these cases. It is commonest, however, for values to be nearer to zero and, for example, Leopold *et al.* (1964, p. 102) quote a range of 0.27 p.p.m. to 6.5 p.p.m. for monthly average composition of rain over continental USA Other figures quoted in the literature range from 0.5 to 4.2 p.p.m. to 6.5 p.p.m. for monthly average composition of rain over continental USA Other figures quoted in the literature range from 0.5 to 4.2 p.p.m. in western Ireland (Gorham, 1957) and a range of 1.0 to 1.5 mg with extremes of up to 5, over the British Isles as a whole (Stevenson, 1968). In terms of total annual input Chapman (1967) records a value of 4.7 kg ha^{-1} a^{-1} in the south of England and Edwards (1973a) records a value equivalent to 17 kg ha^{-1}a^{-1} for East Anglia.

In summary, the order of magnitude involved would appear to be in the region of 0.1 to 10 p.p.m. provided as an atmospheric input, with locally higher values according to marine sources or the addition of limestone dust. Although day-to-day amounts will be small, and variable, the over-all supply per year appears to average out in the order of magnitude of 1 to 20 kg ha^{-1} a^{-1} (but with locally higher figures occurring).

The solute loads of streams draining limestone areas tend to be in the region of 100 to 200 p.p.m. (CaCO$_3$; 40–80 p.p.m. Ca^{2+}) (Smith and Newson, 1974; Sweeting, 1972, pp. 223–7) but for non-limestone areas a range of lower figures is quoted. The mean calcium content of river water in the world is given by Hem (1970, p. 12) as 15 mg l^{-1}. Crickmay (1974) gives 20.5 p.p.m. for the Mississippi, on an average annual basis, while Hem quotes a figure of 42 mg l^{-1} for the same river from 1 October 1962 to 30 September 1963. Figures of 10, 13, and 20 p.p.m. are given as mean values for rivers draining granite, quartzite, and sandstone by Leopold *et al.* (1964, p. 103). It seems that the orders of magnitude involved here are, without a calcium-rich lithology, 1–20 p.p.m. Ca^{2+} and, with calcium-rich rocks present, 50–150 p.p.m. Ca^{2+}.

Since, in areas with limestone bedrock present, a rather special case exists, it is perhaps most useful to consider data from non-limestone environments from which to deduce a quantitative model of calcium balance. Some calculations of calcium budgets do exist in the literature and for a dry heath ecosystem Chapman (1967) suggests the balances shown in Table 7.1. However, this budget does not include leaching losses and this is liable to considerably reduce the gain shown and possibly even give rise to a net loss; this is an unknown in the balance equation and it was not measured. It is demonstrated by Thomas (1969) how the uptake of calcium by dogwood trees (*Cornus florida* L.) acts to minimize the downward leaching of calcium. Figures are given of 12.5 kg ha^{-1} (Ca^{2+}) present in the trees themselves and 170·6 kg ha^{-1} present in the litter. Trees, in general, return about 55–95 per

Table 7.1   *Calcium (Ca^{2+}) balance for a dry heath ecosystem (kg ha^{-1} a^{-1})*

Input (rainfall)	56
Output (burning)	12.5
Gain (12 y)	+ 43.5

Source: Chapman (1967)

cent of their calcium to the ground in litter fall, retaining the rest in their perennial tissues (Thomas, 1969, p. 116; Ovington, 1962). Figures for the calcium of leaves, branches, trunks, and fruits for various tree species are given in Ovington (1962) and include those shown in Table 7.2 (from Ovington, pp. 160–2, Table XV). Between 100 and 600 kg ha^{-1} of calcium may be stored in trees at any one time. Further figures are given for uptake, retention, and release in kg ha^{-1} (from Ovington, p. 174, Table XVI) as shown in Table 7.3.

Table 7.2   *Approximate range of calcium contents of tree stands (kg ha^{-1})*

Pinus sp.	100–200
Picea sp.	200–400
Pseudotsuga sp.	200–600
Betula sp.	300–400
Quercus sp.	150–200
Fagus sp.	*c.* 150

Source: Ovington (1962)

Table 7.3   *Calcium cycling in different vegetation types (kg ha^{-1} a^{-1})*

(a) Vegetation	Uptake	Retained	Released by litter fall
Spruce on podzol	56	8	48
Spruce on rich soil	51	6	46
Birch	107	53	54
Oakwood	102	16	86

(b) A *balance sheet* is given for a *Pinus sylvestris* ecosystem (Ovington, p. 175, Table XVI)

Average uptake	55
Average release by litter decomposition	44
Average loss to store in tree trunk	4
Average change in ecosystem	+ 7

Source: Ovington (1962)

From these figures it can be estimated that vegetation can be responsible for the storage of calcium in the range of 10 to 600 kg ha⁻¹, the precise amount depending on the vegetation type. This should correspond with the uptake of an equivalent amount from the soil. Varying proportions of the uptake may be returned by litter fall and released by decomposition, usually more than half the uptake and often nearly all the uptake (retention in perennial tissues representing a relatively small amount). It is in the litter processes that the balance becomes critical: it is from litter release that calcium could either be leached by drainage waters or taken up again by plant roots. In theory, at least, if litter calcium is leached out of the soil system then further calcium will have to be released by mineral weathering for plant uptake. Conversely, if calcium released from litter is taken up again by roots then it would not be necessary for mineral weathering to occur in order to supply calcium for root uptake. Moreover, drainage waters would have a low concentration of calcium derived from litter decomposition.

The question of relative amounts is important here and it would obviously be useful to compare the amounts of calcium involved in various processes. The questions can be asked: are the amounts of calcium involved in cycling significant relative to the amounts involved in weathering? If all the vegetation were to be removed, would weathering and leaching in fact be affected? In attempting to answer these questions, it is first important to reiterate and re-emphasize the qualifications that comparisons of data are difficult since the units used in various studies are different and it is doubtful whether such comparisons will be wholly valid. However, comparisons will be useful at this stage if only to gain order of magnitude estimates of the amounts involved.

In general, concentrations of calcium in streams and rivers in non-lime-stone areas (1–20 p.p.m.) appear to be equal to or greater than rainfall inputs (0.1–10 p.p.m.) and therefore a mineral weathering supply appears to be implicated as a causative factor in the derivation of calcium in drainage waters. However, deriving budgets for actual amounts involved on a yearly basis and bringing these figures to a kg ha⁻¹ basis is problematical for while a range of 0.01 to 1 kg ha⁻¹ may be reasonably quoted for an over-all atmospheric input, the drainage losses will naturally depend upon drainage density and discharge regimes. However, some data are available on this topic and Edwards (1973a, p. 213) gives figures of 30.2 and 20.4 metric tonnes per km² for 1970 for the Yare and Tud catchments (Norfolk, UK respectively. This corresponds to a range of 300 to 200 kg ha⁻¹ a⁻¹ (it should be borne in mind that these catchments do have substantial amounts of chalky boulder clay in them which acts as a source for calcium). These values are higher than those quoted by Ovington for litter release. It is reasonable to assume that the bulk of this calcium has come from rock weathering directly since the orders of magnitude amounts involved in drainage losses are far in excess of those involved in biological cycling and rainfall input. Whether this would be the case in a non-chalky catchment is difficult to say but it is probable that the amounts involved in leaching loss would be more comparable with those released in litter decay. Since the Norfolk streams involved have a solute load in the region of 150 to 200 p.p.m. $CaCO_3$ then, for a non-chalky stream with a solute load of 5–20 p.p.m., it can be calculated that the average annual losses would be in the order of 20 to 30 kg ha^{-1}a^{-1} for non-chalk streams of similar discharge and regime to the chalk streams. As a rough and speculative estimate, it can be suggested that input by rainfall is of the order of 1 to 20 kg ha^{-1}a^{-1} and output by leaching would be 20 to 30 kg ha^{-1}a^{-1}. Relative to the uptake and release figures from Ovington (1962) (for example, of an oakwood uptake of about 100 kg ha^{-1}a^{-1} and litter release of about 80 to 90 kg ha^{-1}a^{-1}) these figures are small. While this budgeting is speculative it is safe to conclude that input, output, and cycling can operate in such a way as to influence each other. It can be presumed that the bulk of nutrients in vegetation will not be derived from rainfall but by weathering and litter release and that, in a non-calcareous environment, leaching losses could be sensitive to and altered by retention in biological cycles. This is in accordance with the conclusions of Bormann *et al.* (1969) and Redfield (1958).

A question remains as to how much calcium released by litter is recycled by root uptake and how much is lost by leaching in organo-metal compounds. One approach to this would be to study the amounts of calcium which are present in streams in organic compounds.

In limestone areas an increase of calcium in drainage waters in the spring months has been monitored by Pitty (1966). However, this could be due equally to an increase in the metabolic activity of micro-organisms, and a concomitant increase in carbon dioxide available for limestone solution, as much as it could be due to release from the decay of litter and other organic matter stored since the previous autumn and the release of calcium. Mineral calcium carbonate can be solubilized by carbon dioxide (from atmospheric and biogenic sources) dissociated in water to yield hydrogen ions:

$$CO_2 + H_2O \rightarrow H_2CO_3 \rightarrow H^+ + HCO_3^-$$

$$CaCO_3 \rightarrow Ca^{2+} + CO_3^{2-}$$

$$HCO_3^-$$

$$Ca(HCO_3)_2$$

Alternatively, calcium carbonate may be solubilized by H$^+$ derived from organic acids or it may be directly incorporated into organic chelates (see Section 3.3). It is suggested by Smith and Mead (1962) that calcium in combination with organic matter is rare in stream waters in limestone districts but this could be due as much to the fact that at the limestone stream resurgences sampled the organic matter may have already decayed during its journey through the limestone system from the point of origin in the soil/rock layer as to a total absence of organo-metal compounds in the system. More recent work by Bray (1975) suggests that calcium can be transported in organo-metal compounds in streams. However, the work deals with particulate organic matter suspended in streams, together with acids in solution, derived from peat-bog areas rather than with calcium humates derived from the soil zone.

It is, in fact, quite probable that calcium released by litter decay is used by uptake by plants. It is suggested by Newbould (1969) that root uptake chiefly occurs in the upper layers of the soil, where decaying organic matter could be expected to be present. There still remains one point, however, that return does not necessarily equal uptake because of the storage in perennial plant tissues. Excluding leaching loss from litter, net uptake (i.e. uptake minus return) would be equal to storage (for example, in Table 7.3, for oakwood vegetation, net uptake would be equal to 16 kg ha^{-1}, as although 102 kg ha^{-1} are taken up, 86 kg ha^{-1} are returned). This balance would have to be supplied by mineral weathering or rain-water but nevertheless represents only a small fraction of the total amount involved.

It is problematical to find a balance point between leaching, weathering, and cycling. Leaching and weathering could operate more or less independently of cycling, as can be suggested in Fig. 7.8. In this case, only a small amount of the weathered mineral calcium would go into plant circulation. Alternatively, cycling could be involved with weathering and leaching as indicated in Fig. 7.9, in which case weathering inputs would be slightly greater, allowing for some storage in tissues and complete removal of cycled calcium. In truth position is probably between these two extremes, with most of the leached calcium being derived from weathering and most litter calcium being recycled, but precise quantification of the model is difficult because compatible data are lacking. A likely generalized estimate, however, would

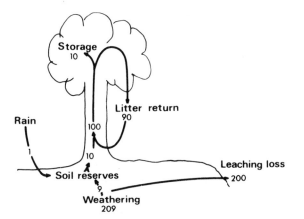

FIG. 7.8. Cycling/leaching model for calcium—alternative 1. Plant uptake from cycling, weathering products leached. Figures in kg ha^{-1}a^{-1} (data from various sources, chiefly Ovington, 1962)

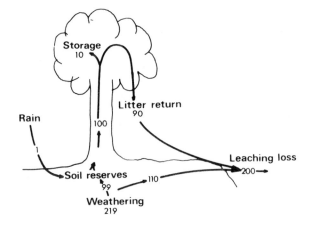

FIG. 7.9. Cycling/leaching model for calcium—alternative 2. Plant uptake from weathering, cycled minerals leached. Figures in kg ha^{-1}a^{-1} (data from various sources, chiefly Ovington, 1962)

be that net uptake by plants is equal to tissue storage and a small amount of litter leaching. In this way there would be small amounts of seasonal variations in calcium in streams, but not either none, or huge, variations as the other extreme models would predict.

If this is so then given static leaching conditions alterations of vegetation type from, say, *Pinus* to *Betula* would act to impoverish the soil simply because birch appears to retain more calcium in its tissues (see Table 7.3(a), as suggested in Fig. 7.10. Similarly, planting spruce would reduce nutrient

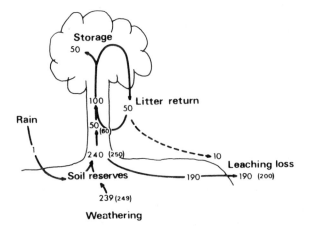

FIG. 7.10. Cycling/leaching model for calcium in *Betula* Figures in kg ha^{-1} a^{-1} (data from Ovington, 1962). An option of 10 kg litter leaching is included

demand and maintain nutrient status. However, these models are too simplistic as the litter decomposition rates and types will radically alter the weathering potential and the leaching loss picture. Oak litter is relatively high in calcium and decomposes rapidly giving a mildly acid humus which encourages microbiological activity, and thus encourages a high release of calcium and efficient nutrient recycling. Birch and especially spruce, however, have a more acid humus, discouraging recycling and encouraging the production of chelates. This will lead to soil impoverishment by increasing the weathering potential and therefore encouraging leaching. Overall nutrient status of the soil-vegetation system is therefore not so much a matter of the absolute amounts of nutrients taken up and returned by recycling, but is more a matter of (1) the nature of the humus and its effect on weathering and (2) the limits placed upon weathering by weathering potential. Therefore models can be made of low and high weathering release. Low release can be due to (1) low potential, even if reserves are high, or (2) low reserves, regardless of potential. In the latter case nutrient uptake and biological production may well be limited. High nutrient release from minerals will be the result of a combination of high reserves and high potential. The planting of plant species with a low return of nutrients in the litter and an acid litter on a calcium-rich soil will result in increased weathering. The planting of these species on a calcium-poor soil will simply lead to a largely unfulfilled rise in weathering potential, i.e. a rise in acidity. In calcium-poor soils it can be suggested that most of the little calcium that is present may be involved in cycling, but since litter decomposition by microorganisms will be limited by lack of nutrients then much of the calcium will be stored in humus. For example, the storage of nutrients in the humus of sand dune soils is recorded by Wright (1955) for conifer plantations on the Culbin Sands, Morayshire. Leaching losses will be low as, although potential is high, there is little weatherable calcium in the soil. Vegetation reserves of calcium will therefore be important in these environments.

In calcium-rich soils a more robust situation exists, weathering can always take place to offset leaching, though this is only up to a certain limit defined by the weathering potential of the soil (and this, in turn, is largely a matter of carbon dioxide productivity). This should be high as in the nutrient-rich soils, soil faunal and microbiological activity can be expected to be high also. High nutrient status systems with a large store of weatherable minerals will therefore be characterized by high microbiologically induced weathering and therefore will be biologically self-reinforcing in terms of nutrient status.

It would seem that the perpetuation of systems at a certain level is inevitable because a series of feedback relationships will tend to act to maintain calcium status at given levels. What is not clear is the interaction of the time scales involved in each reaction. Inevitably, leaching and weathering tend to be far slower processes than nutrient cycling—the lowering of the landscape and the evolution of land-forms take place over hundreds of years, whereas nutrient cycling proceeds in yearly cycles. Taken to an extreme, one might suggest that the alteration of the vegetation could alter the rates of lowering of the landscape! While an extreme statement, there must be some truth in this, though it is doubtful if planting pine trees would change a hill into a plateau in visible time! Clearly, in calcium-rich mineral soils, such as occur in a calcareous bedrock or drift environment, cycling is self-contained and tends to be small relative to the larger weathering and leaching processes. On the other hand, where the geomorphological involvement of calcium is minimal, say on a siliceous sandstone, virtually all the small amount of calcium that will be present will be involved in biological cycling. In a soil/rock system rich in calcium it can be expected that seasonal releases and uptake of calcium will not make their presence felt to any marked degree in stream water quality—this will be governed more by weathering processes such as hydrogen ion supply (which, if biogenic from carbon dioxide, may have some seasonal element present, however) and water flow-through times. However, in areas of low calcium in the soil/rock system then the calcium content of streams can be expected to be far more related to vegetation and recycling processes. Output may be equal to rainfall input, though it may not always be clear if it is a 'straight through' system or if some temporary involvement in cycling has been involved. To this end the paper by Johnson (1971) on the Hubbard Brook studies are of significance. In a discussion on ecosystem geochemistry his writings suggest that the significance of biological cycling relative to chemical weathering and leaching removal can vary markedly, according to the nature of each. He concludes that 'the status of the biological system has a profound effect on the immediate passage of chemicals out of the system' (p. 530). He continues that calcium, being involved in biological activity, is especially sensitive to fixation and release and therefore is unlikely to have passed straight through the system without biological involvement (this is compared with sodium which is less biologically involved and more likely to be independent of variations in biological activity). The conclusion is that the net flux of calcium from a non-limestone system, like Hubbard Brook, is the net output of biological involvement combined with weathering reactions, rather than just being a function of weathering reactions alone. Nutrient cycling is seen as imposed upon an open-ended system of chemical weathering and flux out of the system.

Nutrient cycling can thus be seen as a component part of the movement of calcium from mineral weathering to stream output. But the amounts involved appear to make it likely that while changes of weathering and leaching can affect fundamental changes in vegetation cycling (by affecting vegetation type) changes in nutrient cycling will only make relatively minor changes in weathering and leaching systems—though these changes will be marked, if small. The main changes will be in terms of changes in weathering potential but ultimately the store of weatherable minerals will control the limits of the effects of these changes.

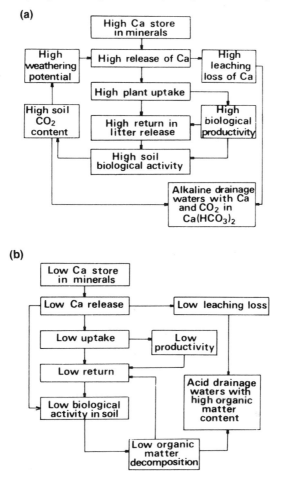

FIG. 7.11.   Functional models for (a) calcium-rich and (b) calcium-poor systems

Thus the chains of cause and effect can be suggested to appear as shown in Fig. 7.11 with (a) a calcium-rich system and (b) a calcium-poor system. Both systems have a high weathering potential, but that of (b) is probably considerably higher than that of (a), possessing not only carbon dioxide but also chelate compounds and dissociated organic acids in greater abundance. However, that of (a) will probably be more readily used up as the system is rich in weatherable calcium.

It is of interest to compare other elements, such as sodium, potassium, and magnesium, with calcium. These, given the limited amount of data available, often seem to have roughly the same order of magnitude of rainfall input and leaching output, though the amounts involved in cycling vary. Summary data from a number of sources (but primarily Ovington, 1962) are shown in Table

Table 7.4 *Approximate order of magnitude estimates of nutrient inputs and outputs for selected mineral cations, with special reference to nutrient systems of temperate forest ecosystems*

Component	Element (kg ha^{-1} a^{-1})			
	Calcium	Sodium	Potassium	Magnesium
Plant Uptake, U:	50–100	1–10	10–100	1–50
Leaching, L:	100–300*	50–100	1–10	1–10
Rainfall input, A:	5–50	1–50	1–10	1–10

*the higher value being for calcium-rich (limestone and chalk) soils.

Source: various, but primarily Ovington (1962)

7.4. In many ways sodium, potassium, and magnesium seem to behave in a similar fashion to calcium, but with a greater or lesser degree of biological involvement. Sodium appears to be less biologically involved than the other three elements.

In summary, in a soil-vegetation system on a parent material rich in calcium carbonate, most of the calcium in the system is involved in weathering and leaching processes. Leaching losses may be as high as 300 kg ha^{-1}a^{-1}. On parent materials poor in calcium carbonate, the amounts of calcium in the system are much less and most will be bound up in organic cycling. If the vegetation is removed most of the calcium in the system will be removed with it. In this situation, rainfall input (of between 5 to 50 kg ha^{-1}a^{-1}) will be the primary source of calcium input. Plants vary considerably in their uptake, storage, and release on litter decay of calcium. In general, however, broad-leaved trees take up and release more than coniferous trees.

7.4.2. Case-study (2): The ecosystem geochemistry of silica

Silica (SiO_2) is present in many different mineralogical forms (Loughnan, 1969, pp. 4–26) but in the context of weathering reactions and plant uptake three main forms can be usefully distinguished. (It should be noted that the term silica refers to SiO_2 while silcon refers solely to Si; thus, technically speaking, silica can also be referred to as silicon dioxide). First, silica can be present as quartz, which consists solely of SiO_2. Second, silicon can be present in various silicate minerals where elements other than silicon can also be present (for example, microcline felspar, $KALSi_3O_8$). Third, it can occur as amorphous silica or silicic colloids, largely in the soil plasma. In runoff waters, silicon can be present in solution as silicic acid, $Si(OH)$, though silicon can also be removed from drainage basins as a sediment in small clay-sized particles of silicate minerals.

The significance of the various forms of silica to weathering reactions is that they can differ markedly in their solubility. The solubility of quartz silica is notably lower than that of most other silicate minerals, especially the amorphous silicates. Elgawhary and Lindsay (1972) give an equilibrium solubility level of 2.8 p.p.m. for Si from quartz and 51 p.p.m. Si for amorphous silica.

Silicic acid is thought to be present in the soil solution in equilibrium with the solid phase (Russell, 1973, p. 635). This is a proposition which is supported by the work of Elgawhary and Lindsay who conducted experiments in which equilibrium was approached both from undersaturation and from supersaturation. In an acid soil they obtained an equilibrium value of 19 p.p.m. Si and in a calcareous soil of 25 p.p.m. Si.

The effect of pH on the solubility of silica is discussed by Loughnan (1969, p. 32). It is suggested that pH values of above 8 markedly increase the amount going into solution (from 3μ mol 1^{-1} at pH 8 to 7μ mol 1^{-1} at pH 10; these data are for amorphous silica). These data are for experimental systems and in natural systems factors other than pH will be important. Therefore it cannot necessarily be expected to gain an exact relationship between pH and silica concentration in natural waters. The presence and concentrations of other ions, and also waterlogging, will be important in affecting silica solubility. The role of other ions is discussed by Acquaye and Tinsley (1965). They demonstrate that the presence of iron or aluminium appears to reduce the solubility of silica in the pH range of 4 to 5 (for iron) and 5 to 6 (for aluminium). The role of humus compounds may also be of significance in altering solubility patterns, either increasing or decreasing solubility. In the experiments by Kerpen and Scharpenseel (1967, p. 222) the addition of humus to the top of leaching columns of three rock types altered the percentage of silica in the leachate—from 53 to 37 (basalt); 52 to 53 (trachyte); 44 to 48 (sand). In natural soil profiles where humus is present together with iron, aluminium, and other ions it is suggested, by Acquaye and Tinsley, that changes in pH (for example, in waters percolating from an acid to an alkaline environment) can effect both the humus compounds and the effects of other ions so that the silica may be precipitated. Organic acids, such as citric acid, increased the solubility of silica in the experiments of Acquaye and Tinsley. For example, using oven-dried samples of a gleyed brown earth, water extractable silica was found to be in the range of 2.1 to 10.9 mg per 100 g soil. Citric acid extractable silica was found to be in the range of 81 to 150 mg per 100 g. Thus, although in pure laboratory solutions the manipulation of pH towards alkalinity results in a greater solubility of silica, this relationship gives a poor prediction of the solubility behaviour of silica in complex, natural situations.

The dissolution processes under alkaline and acid conditions appear to be slightly different. The dissolution of crystalline silica can occur at high pH values with the production of silicate anions or monosilicic acid. Under acid conditions a less direct process is involved. Here the hydrolysis of silicate and aluminosilicate minerals occurs with the replacement of metal cations in the mineral by hydrogen ions from its surrounding acqueous medium. This leads to a disarrangement of the crystal lattice of the mineral and a release of silicon atoms (Acquaye and Tinsley, 1965, pp. 126–7).

A distinction can be made for solubilization in inorganic and organic media. The solubility curves shown by Loughnan (1969) may be theoretically true in inorganic solutions but the provision of hydrolytic hydrogen ions and organic complexing agents under organic, humus-rich conditions can be expected to be responsible for the solubilization of silica in soils; although the effect of organic acids upon silica does not appear to be so marked as it is upon aluminium and iron (as discussed in Section 3.4).

The amounts of silica present in natural runoff waters have been studied by various authors. The figures quoted seems to lie in the range quoted by Elgawhary and Lindsay (1972) of 2.8 to 51 p.p.m., though the upper limit rarely seems to be reached. Flow-through times could be an important factor here. Figures of 17 p.p.m. for groundwater and 14 p.p.m. for stream water are quoted by Davis (1964). Extreme figures for water rich in silica are given by Hem (1970, p. 106) of 103 and 363 μ mol l^{-1} SiO_2 (being from groundwater springs in rhyolite and springs in the Yellowstone Park geyser area respectively). Other figures above the 51 p.p.m. level include 99 μ mol l^{-1} from a flowing well in Idaho, 71 μ mol l^{-1} from a deep well in Idaho, and 62 μ mol l^{-1} from a river supplied by groundwater flow in Nebraska. Of the lower figures quoted by Hem three fall close to the solubility level of amorphous silica (38 and 49 from wells in basalt and 48 for a creek draining extrusive volcanic rocks). A lower figure, 29 μ mol l^{-1}, is given for a well in mica schist. Data quoted by Feth *et al.* (1964) include a value of 25 p.p.m. in SiO_2 in perennial springs and 8 in base flow in streams. These data do seem to suggest that the longer residence time groundflow water has the higher concentrations of silica.

It is of interest to note that Feth *et al.* (1964) suggest that there is very little atmospheric supply of silica. They gained values of 1.7 pp (max), 0.16 (mean), and 0.0 (min.) for snow melt in Sierra Nevada. In support of this Gambell and Fisher (1966), in their study of rainfall in North Carolina and south eastern Virginia, suggest that silica is a negligible constituent of rainwater. Data for silica in rain-water are generally scarce but such data as do exist suggest a source from rock weathering as the primary one for the dissolved silica found in streams. This contrasts with mineral cations such as calcium, magnesium, and potassium which have significant atmosphere sources (see Table 7.4).

Figures available for the leaching losses of silica in kg ha^{-1} vary from the results of one research worker to another and thus, probably, from one area or set of conditions to another. Drainage losses have been calculated by Kerpen and Scharpenseel (1967, p. 218) as being in the range of 15 to 34 kg $ha^{-1} a^{-1}$ SiO_2, the range being composed of the values shown in Table 7.5.

Figures are given by Russell (1973, p. 699) for losses of silicon in rivers.

Table 7.5 *Annual silica losses from various soil types (kg ha^{-1})*

Rendzina soils	28
Brown earth	15–34
Acid brown earth	28
Podzol	34

Source: after Kerpen and Scharpenseel (1967)

These figures can be converted to silica, and work out at around 30 kg ha$^-$ for North American and European rivers and for the Amazon basin. A result of about 25 kg ha^{-1} loss was obtained for a plot of land at Rothamsted Experimental Station.

Figures for concentrations in surface streams are usually less than figures given for concentrations on waters draining from soils or leaching columns (from various sources the former would be in the range of 10 to 15 p.p.m. and the latter would be in the range of 20 to 30 p.p.m. SiO_2). Two main points are relevant here. First, although the concentrations in streams are lower, discharges of water involved are far greater than the discharges of soil water and therefore the total amount removed will be far greater than the values for concentration would seem to indicate (see Section 5.3). Second, it could be that plant uptake and retention play an important role in influencing the concentrations of silica found in streams. Unlike other cations it seems that silica may be returned in plant litter to the soil in a form that is not necessarily readily soluble. Silica is persistent in plant tissues in microscopic opaline particles which are termed phytoliths. It is suggested by Acquaye and Tinsley (1965) that these accumulate as the litter organic matter decomposes and in this way the phytoliths contribute to the fine mineral fraction of the soil. They suggest that as much as 40 kg ha^{-1} a^{-1} may be contributed in this manner. This phenomena is of interest in that the weathering of primary silicate minerals by biological weathering and plant uptake would not necessarily find any expression in the levels of silica to be found in streams. The solubility of phytoliths appears to be greater than that of primary silicate minerals but less than that of amorphous silica (Fig. 7.12(a)). Thus there are two possible roles of phytoliths in silica mobility. First, silicate minerals may be solubilized in the rhizosphere and silica may be returned to the soil in phytoliths which will be more soluble than the primary minerals. Mobilization and leaching loss will thus be aided (Fig. 7.12(b)). Second, uptake from amorphous silica could result in the deposition of phytoliths which would be less soluble than the original form (Fig. 7.12(c)).

In this context some useful experiments have been undertaken by Lovering and Engel (1967) on the uptake and storage of silica by plants. Initially, they investigated two possibilities. The first was that silica was taken up by the plant from silica which was already present in the soil solution. The second

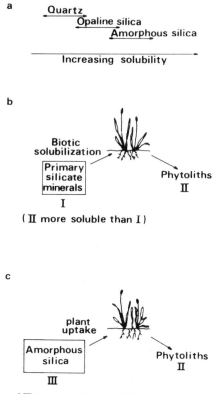

FIG. 7.12.   The possible roles of phytoliths (opaline silica) in silica cycling. (a) Relative solubility (after Verstraten, personal communication). (b) Plant uptake from primary silicate minerals. (c) Plant uptake from amorphous silica

was that uptake was from silica which had been solubilized from primary minerals by plant roots. It was demonstrated by experimental procedures that biochemical reactions taking place at the roots solubilized the unweathered rock with which the roots were in contact. The silica was then transported as part of an organic complex into the plant. Symbiotic microbes in the rhizosphere were implicated as the main agent providing weathering potential. Suggested forms of organo-silicate compounds found in grass plants are shown in Fig. 7.13. Some plants have higher uptakes and contents of silica than others (Russell, 1973, pp. 634–9). Rodin and Bazilevich (1967) give values of 0.01 to 2.74 per cent (dry weight) for deciduous trees and 0.08 to 5.05 per cent (dry weight) for steppe plants. Characteristically, members of the *Graminae* have the highest uptake of all perennial plants. Lovering and Engel (1967) calculated that the uptake of silica would be in the region of

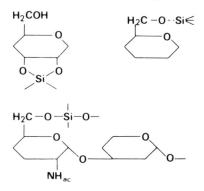

FIG. 7.13.   Some suggested forms of organo-silicates found in grass plants (from Lovering and Engel, 1967)

140 kg ha^{-1} a^{-1} for the whole plant and, in their experiments, losses by leaching would have been equal to 20 kg ha^{-1} a^{-1}. If uptake is as high as suggested, and leaching losses as comparatively low as suggested, then the proposition of Acquaye and Tinsley (1965) that significant proportions of silica are retained in phytoliths is supported. In the Hubbard Brook catchment Bormann and Likens (1969) again give a figure for leaching loss which is low when compared with the figures for plant uptake obtained above. They quote a total of 19.6 kg ha^{-1} lost per annum (3.69 as inorganic particulate matter—especially clay material in suspension—and 15.92 in solution). While the figures given are not strictly comparable because they come from different areas and regimes, it does seem, however, that retention of biologically solubilized silica is a significant factor in reducing the silicon content of streams.

In summary, in modelling the ecosystem geochemistry of silica in a general way, there are several general propositions to be made. First, the mineral form can substantially alter the solubility of silica; second, for any given mineral form, the presence of humus can also substantially alter the solubility of silica. Plant roots appear to be able to effect much of the solubilization that takes place in the soil. Plant uptake of silica cannot necessarily be equated with the return of equally soluble silica in plant litter; microscopic phytoliths of silica are released and may accumulate in the soil (unless themselves solubilized). There thus may well be a discrepancy between uptake of silica by plants (of up to 150 kg ha^{-1} a^{-1} and losses in drainage water (10–15 kg ha^{-1} a^{-1}). Studies of the amounts of silica solubilized from soil columns (15–30 kg ha^{-1} a^{-1}) need not necessarily predict the silica content of streams because of (a) plant retention and of (b) dilution by non-silica-rich waters. Silica values in streams seem to be in the range of 10 to 15 p.p.m., but locally they may rise to much higher figures, especially in groundwater where

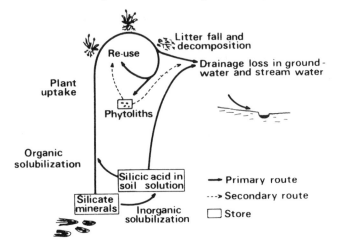

Fig. 7.14.   General model for the ecosystem geochemistry of silica

longer residence times will prolong the contact of water with silica-bearing rocks. An over-all, summary model is suggested in Fig. 7.14.

Further reading

Ecosystem processes. (Jordan, 1985; Kittredge, 1948; Reichle, 1970; Sopper and Lull; 1967; van Dyne, 1969; Waring and Schlesinger, 1985.)

8 Models of Stability and Change in Whole Systems

Man masters nature not by force but by understanding.

BRONOWSKI

8.1. The purposes of studying stability and change

Soil and vegetation systems are used by man as a resource in many ways. They are used in an active sense in terms of agricultural production and they have many uses in a more passive sense—for example, as an influence upon water resources and as a background for recreation activities (see, for example, Curtis *et al.*, 1976, Chs. 14–16 for a discussion and examples). In all cases the systems are being altered and influenced. It is therefore of crucial importance to try to evaluate what effects man has upon soil and vegetation systems. Primarily, the concern is with how far man can apply pressures to a system before the pressures cause an alteration to a new and different state. If it is clear that alteration is an inevitable consequence of a certain action, or group of actions, then value-judgements must be considered and a decision taken as to the desirability of the new state. Thus the environmental scientist should be concerned with the evaluation of the resilience of natural systems and the identification of thresholds, which, if crossed, will lead to the occurrence of a non-returnable (irreversible) state. If these concepts are known it will be possible to evaluate the desirability of a state in terms of its stability in the face of the forces acting upon it.

8.2. Fundamental concepts of stability and change

The various components of soil and vegetation systems, and the various combinations of these components, will display differing degrees of resilience in the face of disturbance. In many ways these degrees of resilience are related to the amounts of energy which are flowing through the system relative to the degree of disturbance to the system. The topic of disturbance and resilience of ecosystems is reviewed by Pickett and White (1985), and also by Waring and Schlesinger (1985). These authors stress the need for an integrated approach to soil and vegetation systems if disturbances and recovery are to be understood; especially important topics are the sensitivity of the components, community dynamics, energy flows, and nutrient cycles.

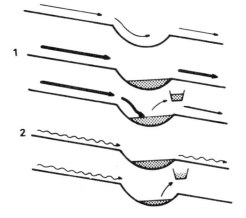

FIG. 8.1.   Disturbance and recovery—an analogy of stream inflow, pond levels, and water removal in buckets

The concepts of disturbance and resilience involved can be most easily illustrated by analogy. If a bucket is dipped into a pond which is being fed by a stream and if the bucketful of water is removed from the pond, then water immediately flows from the stream into the pond to replace the water extracted. The rapidity of replacement will, however, depend upon the amount of water flowing in. If, instead of a stream; there is a mere trickle feeding the pond then the quantity removed in the bucket will be large when compared to the input flow of water. Thus removal of a bucketful of water will cause a temporary lowering of the level of the pond until the trickle manages to replace the extracted water (Fig. 8.1).

Two extreme cases of his analogy can be envisaged. On the one hand, a large river can be imagined where the extraction of a bucket of water will be substantially immaterial to the state of the river, and on the other hand, a small, stagnant puddle can be envisaged where the removal of a bucket of water will cause the permanent lowering of the water level (until, perhaps, an ensuing fall of rain fills it up again). These cases are analogous to robust and delicate soil and vegetation systems respectively and the former can obviously resist more external pressures than the latter.

Thus, by an analogy, it can be demonstrated that different natural systems have different degrees of resilience and recovery in the face of disturbance. The resilience depends upon the amount of input into the system relative to the amount of the disturbance.

The rate of input becomes even more crucial if the disturbance is a repeated one. If twenty buckets of water are removed from a system with a rapid and large input, the system will still maintain its form, i.e. the water level will be maintained. Conversely, with a low input the repeated removal of water will soon exhaust the store of water and alter the state of the system. In this latter case subsequent recovery will be dependent upon the rate of input (assuming that the disturbance ceases once the store has been exhausted). If the potential for disturbance is always present, i.e. if water is extracted as soon as it collects, the form may never recover. Moreover, if the input is intermittent, like the rainfall into the puddle, then the system will be even more susceptible to the effects of the disturbance (assuming that the disturbances are more frequent than the input events).

It is clear that the crucial factor which must be identified in the study of the stability of natural systems is the rate at which the disturbance is acting relative to the rate at which the system can recover. Thus stability cannot be given any general, objective definition but only definition in the face of the forces acting upon it. Given identical forces, the most stable system will be the one with the greatest amount of replenishing, input energy. Given identical inputs the most stable system will be the one which is subject to least disturbances. It follows that if a general stability model, in the form of a diagram, is to be proposed it must be based upon the degree of disturbance relative to the resistance of the system (Fig. 8.2). Simply, a system can equally well be stable because of low resistance and low disturbance or because of high resistance and high disturbance. But if disturbance gains the upper hand the balance point is passed and the system will be located in the bottom right-hand sector of the diagram. The most stable system is clearly that in the top left-hand sector, with high resistance and low disturbance.

A discussion of the magnitude and frequency of disturbances relative to the magnitude and frequency of recovery inputs forms the basis of this chapter. Clearly, the aim should be to characterize a system according to its position on Fig. 8.2 and to attempt to identify the balance points involved. However, there is one further fundamental factor which has yet to be considered. This is the factor of system reaction.

While a disturbance to a system may act to affect the state of the system not all systems receive the disturbance passively. A changed state of a system can, in certain circumstances, affect the nature of the input. This is not the case with the pond and bucket analogy since the inputs are external and independent of the pond system. However, in some systems, input and the state of the system may be mutually reactive. Hence a lowering of the status

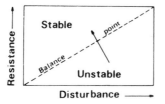

Fig. 8.2.   Diagrammatic representation of stability as a function of resistance and disturbance

of the system by disturbance may act to renew the vigour of the input so that it attains a level higher than that at which it normally operates. Because of this type of feedback effect a system under stress may be more resilient than would be indicated by a study of the system inputs under normal conditions.

The feedback type of reaction can be illustrated by a simple, everyday analogy. If a person experiences a loss of heat by going outside he simply puts on protective clothing. Thus the effect of the cold is not as marked as it might have been if he had just kept his normal, indoor clothes on. A more sophisticated analogy is that of a thermostat, as described by Chorley (1962). If an oven cools down past a certain threshold of temperature the thermostat switches the oven on again so that the level of heat in the oven is maintained.

An environmental example can be given by envisaging the removal of vegetation from an area. Vegetation will soon recolonize the area, either by more rapid growth of existing species of plants or by the invasion of rapidly growing pioneer species. In this way the vegetation cover will re-establish itself by reacting to clearance with more rapid growth rates than existed prior to clearance.

The effects of disturbances, then, can be offset by reactions to the disturbance (by the mechanisms illustrated). If the reactions are strong enough, they may act to maintain the state of the system at or near the level that existed prior to the disturbance, even in the face of the disturbance. Figure 8.2 can now be redrawn to illustrate the three-dimensional stability field that exists when degree of disturbance, degree of input, and degree of reaction are all considered (Fig. 8.3).

In summary, it will be important to discuss two themes in any consideration of system stability. These are first, the nature of the system input relative to the nature of the disturbance and second, the degree of reaction which occurs in the system in response to the disturbance.

8.3. Problems of application

Soil and vegetation systems are complex, multi-component systems. One of

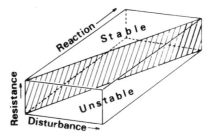

FIG. 8.3.   The stability model with system reaction included as well as resistance and disturbance

the major problems in the application of stability concepts is that some components will be more changeable than others. Thus, if there is a change in external conditions, some components will adjust more rapidly than others. On a very broad level, the major differences between the abiotic and biotic components will immediately be apparent. For example, if a stand of vegetation is removed, natural re-seeding and regrowth will quickly re-establish a vegetation cover. If a hole is dug in the soil it will take a very much longer time period before soil will re-form.

In this way a hierarchy of responses could be envisaged. Were climate to change then it would be expected that first vegetation would change in response, then soil, and lastly rock weathering. But this occurrence of a simple progression from vegetation to soil to rock response is not necessarily the case. It would only apply in certain circumstances where, for example, a decrease in temperature might occur. Here, plant growth would slow down and the rate of organic matter decomposition would decrease. The effect on the soil would be secondary and consequent upon the vegetation change. If, on the other hand, an increase in rainfall were considered it is probable that the leaching of soils would increase and that a vegetation change would be consequent upon a soil change. Different response chains will thus be seen with different external changes.

In the application of the fundamental concepts it can be seen that it is very necessary to specify the precise situations and factors which are being dealt with. This is because the various components of the system operate on different time scales and because different external factors place stresses upon different causal links in the chains of interrelationships between system components.

There is also the relationship between the nature of the components and their responses to disturbances to consider. Some components may be very prone to change, and these can be termed *labile*. Others may be less responsive and these display *low lability* or are *non-labile*. Because of the multi-component nature of soil and vegetation systems, and the different lability of the components, the true position of disturbance and recovery is not as simple as shown in Figs. 8.2 and 8.3. There may be only one external disturbance but the system may well be composed of multiple input and reaction types and therefore display a heterogeneity of responses to a single disturbance. Heterogeneous lability of multiple components will thus be the key of complex soil and vegetation systems and, as such, their behaviour will be difficult to predict. However, the position can be elucidated if the reasons for lability or non-lability of each component are understood.

Components can be non-labile for two, very different, reasons. Either they may display a high degree of energy input (relative to a disturbance) and so they can offset the effects of the disturbance and remain in a constant state over time or they can be immune to the operation of the disturbance. In the latter case this may be because the system has been degraded so far that it has not sufficient energy or input power or reaction to offset the effects of the disturbance or it may be because the disturbance is irrelevant to the workings of the system.

For example, a vegetation cover growing at a rate of 1 cm per year will be non-labile in the face of trampling pressure acting to remove 1 cm per year. The form of the cover will remain constant over time because of a constant growth input. The rock weathering processes beneath the grass will be non-labile in the face of trampling disturbance because the operation of the pressure, in itself, is largely irrelevant to it. The system would also be non-labile in the face of trampling if previous trampling had already removed the vegetation cover to leave bare rock. It can be added that obviously a vegetation cover growing at 0.5 cm per year will be labile in the face of trampling pressures acting to remove 1 cm per year.

The effects of trampling on a chalk grassland are described by Chappell *et al.* (1971) while Liddle (1975) gives a broader, more theoretical view of vegetation disturbance and recovery. In the latter work, resilience, or degree of input, is equated to primary productivity. Disturbance is measured by the number of passages of feet along a single file route over the vegetation which reduced the vegetation cover or biomass to 50 per cent of its original state. Although several sources were used in the work, and there were clearly difficulties in using data which were not necessarily compatible, a general relationship between productivity and resistance to disturbance was worked out. The relationship can be expressed:

$$\log_{10} \text{(number of passes)} = 1.362 \log_{10} \text{(productivity)} - 0.896$$

The relationship is illustrated in Fig. 8.4. The vegetation types are numbered and they are, in order of increasing resilience:

1. Snow bank community.
2. Woodland ground flora (chiefly *Vaccinium myrtillus*).
3. Stone stripe community.
4. Acid heath.
5. *Callunetum*.
6. *Ammophiletum*.
7. Dune grassland.
8. Sand dune pasture (winter).
9. Sand dune pasture (summer).

A basic criticism of the graph is that the time intervals between trampling were not necessarily comparable. This is an important point, since recovery will clearly depend upon the frequency of the disturbance as well as on primary productivity. However, the general relationship is still valid because, even if the data derived from very short time intervals of passes are ignored (observations 2 and 3) leaving only the more comparable data, the general trend of the data may still be observed. As well as productivity, the factor of

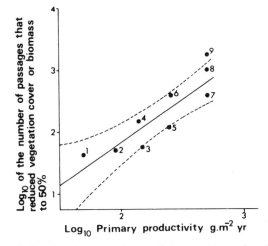

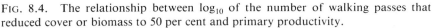

FIG. 8.4. The relationship between $\log_{10}$ of the number of walking passes that reduced cover or biomass to 50 per cent and primary productivity.

The numbered points refer to vegetation communities listed in the text. Calculated regression: —————————; 95 per cent confidence limits - - - - - - - - (from Liddle, 1975)

plant life-form is important. Small wiry plants with basal meristems (growing points) will tolerate trampling better than any larger, taller or soft, fleshy plants. Moreover, the system itself may react to pressure, in that under light pressure, species diversity and productivity may increase; only to decrease when trampling is severe. Plant communities may adapt to pressure so that plants with high potential productivity become common on paths.

In this actual example, then, the disturbance/recovery relationship can be understood in a general way in terms of productivity. In specific situations, plant types and reactions to pressures become important.

The questions of reaction and thresholds are extremely important. Thresholds operate in one of two ways. The first is when the passing of a threshold acts to accelerate and reinforce the effect of the disturbance and the second is when the passing of a threshold acts to diminish the effects of a disturbance. These response types are examples of positive and negative feedback respectively. The thresholds involved can be termed irreversible and reactive.

The case of an irreversible threshold can be illustrated by the example of grazing pressure. Up to the point when the threshold is reached the system will be able to offset the effects of the pressure by energy input and reaction to the disturbance. But beyond the threshold the capacity for renewal would be exceeded. Thus if grazing pressure was to be increased the vegetation cover would become degraded. In practice, however, grazing pressure may be complicated and variable over time. An interesting study of grassland changes with changes in rabbit grazing pressure (as a result of myxomatosis) has been made by Thomas (1963). Vegetaion responded most closely to change in pressure but some soil properties also varied.

In the case of a reactive threshold, the system may not offset the disturbance in any way until a threshold is reached. The passing of the threshold may trigger a response which would then act to offset the effects of the disturbance, either returning it to its former level or maintaining it at a new, but lower, level. For example, a vegetation cover may be degraded in an area by grazing until bare soil patches appear. At this stage new species may invade an area and these may have a more active or robust growth, thus acting to return the vegetation cover towards its former level.

Irreversibility or delayed reaction may be seen without the presence of thresholds. A system could gradually deteriorate in the face of constant pressure or in the face of increased pressure. Deterioration would begin at the inception of the pressure rather than after a threshold has been passed. In these cases it would be clear that the system had no resilience in the face of the pressure or disturbbance. Similarly, the system could react progressively to the disturbance, the reaction acting to maintain it at a given level.

In summary, the application of the fundamental concepts of stability to soil and vegetation systems meets with certain problems. These arise largely from the multi-component nature of the systems. Each component will have a different lability and each disturbance may stress different causal links that exist between the components. Moreover, it may be necessary to identify thresholds, over which the relationships between disturbances and the system inputs and responses may alter.

8.4. General models of disturbance and recovery

Generalized, theoretical models will be discussed first (which could have application in a number of contexts) and their application to soil and vegetation systems will be considered subsequently in Section 8.5.

The state of a system can be modelled over time and in the face of disturbances. In the modelling which follows the state of the system is indicated in diagrammatic form by a vertical axis. The progression of time is indicated on a horizontal axis from left to right. Disturbance events are indicated by a vertical arrow operating from above. Recovery of a system is indicated by a dashed line. The general diagram is illustrated in Fig. 8.5. In the modelling the reasons for the patterns are not discussed: the reasons will depend upon the nature of the individual system, as will be discussed in Sections 8.5 and 8.6. Many general models are suggested below (and there may indeed be other variants). The intention is to see whether the models can be usefully applied to soil and vegetation systems. If they can be it is hoped that it will be possible

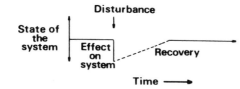

FIG. 8.5.   General disturbance and recovery model

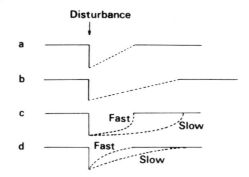

FIG. 8.6.   Different recovery types, single disturbance. (a) Rapid linear recovery (b) Slow linear recovery (c) Cumulative recovery (d) Exponential recovery

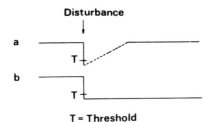

T = Threshold

FIG. 8.7.   The operation of thresholds. (a) Reaction and recovery after threshold is passed (b) Irreversible threshold

to say how far pattern recognition will be useful in the prediction of the future trends and stability of the systems.

Having illustrated the basic model in Fig. 8.5, the variants are illustrated in subsequent figures. The details of the models are explained in the Figure captions but Figs. 8.6, 8.7, and 8.8 deal with single disturbance models (where T = threshold), and Figs. 8.9 to 8.12 with multiple disturbance models, and Figs. 8.13 and 8.14 with continuous disturbance models. Lastly, Fig. 8.15 deals with systems which are already prograding or degrading prior to disturbance.

FIG. 8.8.   Models of immune and nearly insensitive systems

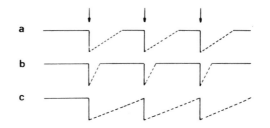

FIG. 8.9.   Multiple disturbances—recovery rate is greater than the frequency and effect of disturbances; with various rates of recovery, (c) is more finely balanced than (b) and is potentially more sensitive to disturbances

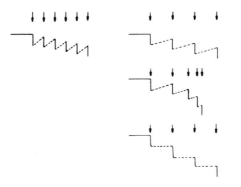

FIG. 8.10.   Multiple disturbances—recovery rate is less than the frequency of disturbances, no feedback (with variants)

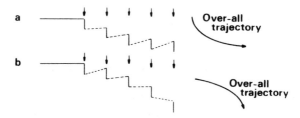

FIG. 8.11.   Multiple disturbances. (a) Negative feedback (b) Positive feedback

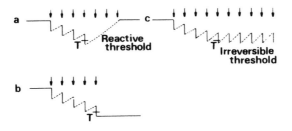

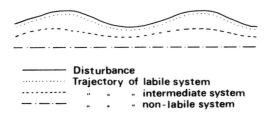

FIG. 8.12. Multiple disturbances with thresholds. (a) Reactive threshold returns system to its prior level. (b) Reactive threshold maintains system at a new lower level. (c) Irreversible threshold

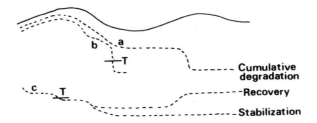

- ——————— **Disturbance**
- ············ **Trajectory of labile system**
- — — — — —   "   "   " **intermediate system**
- — · — · —   "   ·   " **non-labile system**

FIG. 8.13. Continuous disturbance model

Disturbance is represented by a fluctuating line rather than a series of arrows. Some factors will be sensitive to continuous disturbance and others will be less sensitive

FIG. 8.14. In a labile system the disturbance may have a cumulative effect (a) or an irreversible threshold may be passed as in (b). Recovery and stabilization are shown in (c)

8.5 Application of general models of disturbance and recovery

Having illustrated the general model (Fig. 8.5) and the theoretical combinations of disturbance and recovery types (Figs. 8.6 to 8.15) it will now be appropriate to consider how far these can be used to elucidate stability and change in soil and vegetation systems. This can be usefully effected on three levels. First, the disturbance and recovery models can be applied to each separate causal link between components in turn. Then, second, an overall

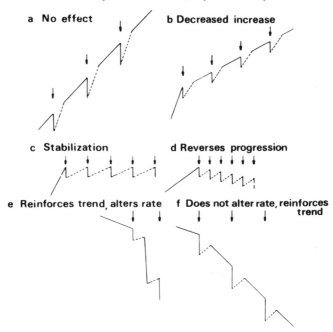

FIG. 8.15.   Effect of disturbance on already prograding or degrading systems

The system is developing to a new state prior to the disturbance. The disturbance may act against the prior trend (a)–(d) or it may reinforce it (e), (f). (a) The disturbance has little effect on the progression of the system. (b) The progression is slowed down. (c) The progression is halted and the system is maintained at a new level. (d) The progression is reversed. (e) The trend is reinforced by the disturbance of the rates of operation. (f) The trend is reinforced by the curtailing of the recovery

view can be taken and it can be assessed how far the alteration of one component or link will act to influence the disturbance and recovery of the whole system. In both the first and second cases internal links, processes, and mutual adjustments can be discussed. On the third level, the external relations of soil and vegetation systems can be discussed with reference to anthropogenic pressures and disturbances. This evaluation will be based on a knowledge of the internal sensitivity of the system and it will be possible to assess how far human activities will ramify through the system. Clearly, some human activities can cause widespread effects and promote a series of chain reactions. The effects of others may be quickly dampened by the system. The challenge is to attempt to identify the factors which control the nature and extent of the effects of the disturbances.

   The application of disturbance and recovery models will become complex unless an overall framework is established first. Nutrient systems will be used as a starting-point and they will be discussed in the overall framework of

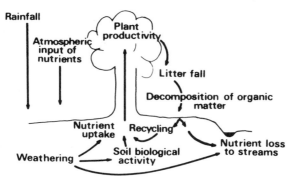

FIG. 8.16.   Nutrient system components

ecosystem geochemistry. The ecosystem geochemistry model has been discussed in earlier chapters but it will be useful to reiterate it here in diagrammatic form (Fig. 8.16). Using the diagram each link between components discussed below can be identified and located in the overall scheme.

8.5.1. Disturbance and recovery in nutrient systems

The nutrient status of a soil and vegetation system is strongly influenced by the degree of chemical weathering in the soil. As has been discussed earlier, rainfall events, losses in drainage waters, and biological activity will be the chief controlling factors acting upon weathering input into the system. It will thus be relevant to model the changes in nutrient status over time in accordance with the relationships between weathering, rainfall, drainage loss, and biological activity. The store, or reserve, of nutrients in the rock will also have to be considered. As in earlier chapters the focus of attention is the nutrient mineral cations.

8.5.2. Abiotic model—the effects of rainfall and drainage

As a starting-point the effect of drainage output upon nutrient status will be considered. The progression of nutrient status without nutrient output is shown in Fig. 8.17. In this diagram and the ones which follow, the vertical axis represents the nutrient status of the whole (soil + vegetation) system.

FIG. 8.17.   Nutrient systems over time, weathering input

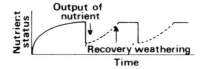

FIG. 8.18.   The effect of output disturbances

Output of nutrients in drainage waters represents a disturbance to the nutrient status of the system. The depletion of the system of nutrients should give rise to an increase in weathering potential (see Fig. 8.26, below). Assuming that nutrients are present in a weatherable form in the mineral portion of the system then further nutrients will be released by 'recovery' weathering to offset those nutrients lost (Fig. 8.18).

In a steady state system, drainage loss output would equal weathering input and the progression of the system through time would be constant. If, for any reason, the outputs and inputs were imbalanced the pattern may be seen as shown in Fig. 8.19. A downward trajectory (Fig. 8.19(a)) would simply mean that weathering input rates were less than drainage loss amounts. An upward trajectory (Fig. 8.19(b)) would indicate that the weathering input was greater than drainage output. It is difficult, in fact, to see how either of these could continue in their directions indefinitely, especially the latter trajectory. Rates of weathering depend upon the supply of weathering potential as much as they do upon the supply of nutrients in the rock and on output. Thus the downward trend would tend to stabilize. This is because weathering potential would greatly increase as the nutrient status became lower (assuming the presence of weatherable minerals). In the absence of weatherable minerals the status would stabilize at the level of supply of nutrients in rain-water and atmospheric fallout. Similarly, the upward trend would continue until a lack of weathering potential lead to a decrease in weathering input; the weathering potential being constrained by a lack of output and an accumulation of weathering products. The stabilization of a downward trend is shown in Fig. 8.20(a) and of an upward trend in Fig. 8.20(b).

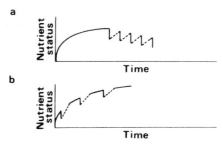

FIG. 8.19.   Output and input imbalance

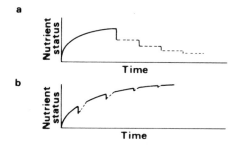

FIG. 8.20.   Stabilization of (a) a downward and (b) an upward trend

Both these stabilization levels represent equilibrium states—where the inputs and outputs are balanced. It is of interest to differentiate between the two equilibrium states, however, because they differ in their potential stability.

One equilibrium state can be recognized where the nutrient status is constant because the mineral supply is exhausted (Fig. 8.20(a)). The other equilibrium state can be recognized where the nutrient status is constant because of the operation of a factor acting to maintain it at a high level (Fig. 8.20(b)). These two states will show differing responses to external changes in various factors.

For example, the nutrient status of a soil could be maintained at a high level because of poor drainage. The status is poised at a high level and the artificial improvement of drainage could soon lead to a marked loss in nutrients. The equilibrium is potentially unstable. On the other hand, if the nutrient status is at a constant low level because the nutrient supply of the mineral portion of the system has been exhausted then alteration of drainage will have very little effect upon nutrient status. The equilibrium is a stable one.

The differences in equilibria are analagous to the positions of two footballs, one in a valley bottom and one on a mountain top. The former, however much it is kicked around, will return to its stable position in the valley bottom. The one on the mountain top, however, is potentially unstable. If dislodged, and the forces holding it up are thereby removed, then it will roll rapidly downhill to a new position. This is a crucial differentiation of equilibrium states. It is not sufficient simply to identify an equilibrium state as one where no changes occur over time. It is vital to identify the forces which are keeping the system in that state and then to evaluate how labile those forces are.

In soil and vegetation systems it is possible to identify both passive and active forces which maintain equilibria. The passive forces are ones like poor drainage which can be altered and where there is no reaction mechanism to change the system back to its original state. The active ones are largely concerned with the flow of energy and matter. If altered as in the case of the stream and the bucket of water described above they will quickly react to maintain their foimer level. These reactive systems are more likely to exhibit stable equilibrium characteristics and recovery after disturbance, albeit that they will probably have a threshold over which recovery will be impaired or absent.

The complicating factor in the study of the disturbance and recovery in nutrient systems is that they are composed of both active and passive stabilization components. Moreover, the active components may each react on different time scales. Thus, if an abiotic component is stressed or altered, it can be expected that a different reaction will occur than if a biotic component is stressed.

It will be useful to recap the foregoing discussion before continuing. Abiotically speaking, a system will stabilize because of constraints on input (lack of weatherable minerals) or constraints on output (drainage). In the former case the nutrient status will tend to become minimal over time. Alteration of rainfall amount and drainage amount will have little effect upon the nutrient status of the system since they are quantitatively unimportant in maintaining it. In the latter case there exists a potentially unstable system at equilibrium and the factors acting to maintain the equilibrium may be changed and the equilibrium will be altered.

To continue, it will now be of interest to examine the stability of the minimum nutrient status system rather more closely. Its stability is not as simple as might have appeared above. It is true that it is stable in the face of drainage changes but it is not stable in the face of changes in nutrient supply from the atmosphere.

Since a minimum nutrient status system is dependent upon nutrient supply by rainfall and atmospheric fallout the status will be very sensitive to fluctuations in nutrient supply from these sources. This is illustrated in Fig. 8.21. The loss of weatherable nutrients from the system and stabilization is shown (and this applies equally to the leaching of a quartziferous sandstone or to the growth of a peat bog which will effectively isolate the mineral portion of the

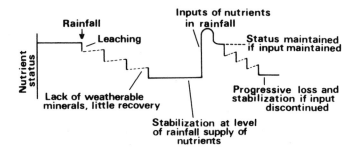

FIG. 8.21.   Disturbance and recovery in a nutrient-poor system, effect of variations in the nutrient content of rainfall

system from the rest of the system). Disturbance may take the form of an input of nutrients in the rainfall and the recovery then takes the form of a leaching loss of these nutrients and return to the former, lower level. The nutrient status may be maintained at a level where input equals output if the supply is continued.

It can be seen that the stable equilibrium of the minimal nutrient system only occurs in the face of alterations to drainage. The equilibrium nutrient status is far from stable if the rainfall input is altered. Thus the stability is conditional upon the specification of the factor involved. As discussed before there is no absolute, objective definition of stability, there is only stability relative to specified factors. The term *conditional stability* might be usefully used in this context. Moreover, if a factor, like drainage, is of no importance in maintaining the equilibrium state then it cannot be expected that a change in it will have any effect on the system. On the other hand, if a factor is important in maintaining the system it can be expected that changes in it will affect the system. This conclusion can be expressed in a general way: *the sensitivity of a state of a system to fluctuations in external factors increases in proportion to the importance of these factors in maintaining that state.*

This principle can be illustrated by the fact that the converse of responses is true in systems where the nutrient status is maintained at a high level by poor drainage. Marked fluctuations may occur in the nutrient input in rainfall but the overall effect on the system will be slight because the rainfall nutrient input is quantitatively unimportant relative to the amount of nutrients already in the system. However, the improvement of drainage will result in a lowering of nutrient status, as shown in Fig. 8.22(a). If a threshold is passed and drainage acts to increase weathering potential then further weathering can be expected to release more nutrients and maintain the status of the system at a moderate level, as shown in Fig. 8.22(b).

Of course, the quantitative importance of factors can change if an otherwise unimportant factor changes catastrophically. If rainfall chemistry changes dramatically the system will become sensitive to fluctuations in it, but only in proportion to the quantitative significance of the supply. Thus, if 80 per cent of the nutrients came from source *A* (for example, mineral weathering) and 20 per cent came from source *B* (for example, rainfall

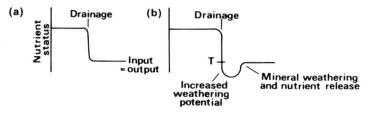

FIG. 8.22.   Drainage and nutrient status

solutes) and if *A* fluctuated widely, the whole system would alter; if *B* fluctuated widely there would be little alteration of the system. However, if the fluctuations of *B* included an increase so that its relative contribution became more than the amount given by *A*, then, clearly, the system would be sensitive to fluctuations in *B*. Thus, while the response to fluctuations will be proportional to the significance of the contribution, there will also be a sliding scale built into the system to allow for the fact that fluctuations may well also include a variation in the proportion of the contribution.

The last abiotic component variable which will be considered is fluctuations in rainfall amount. Similar patterns will be seen for increased rainfall amount as for increased drainage (Fig. 8.22). Increased rainfall amount results in an increased solubilization of the more rapidly dissolving minerals (because of more rapid throughflow, as discussed in Chapter 5). It also results in increased weathering potential. Since weathering potential is also partly a biological phenomenon it will be appropriate to bring in biological factors at this juncture.

8.5.3. Biotic model—effects of biological activity on abiotic model

The patterns described above in Section 8.5.2 may be modified, to a greater or lesser extent, by biological activity (Swank, 1986). There are two principle factors to consider: biological cycling of chemical elements and the effects of the decomposition of organic matter. The former will principally act in two ways—to affect weathering potential and to affect cation exchange. Furthermore, within given limits, the system can react to sustain given levels of nutrient status. Beyond these limits, vegetation type or humus types may change to a different form.

Basically, release and uptake from biological components in soil and vegetation systems will be able to mollify the effects of fluctuations in some external conditions, like, for example, rainfall and drainage. Retention by nutrient cycling and absorption of cations onto organic matter will act to offset drainage losses. In addition, the decay of organic matter releases weathering potential which should release further mineral nutrients to offset nutrient losses. Stability should be achieved when weathering potential, weathering release, biological cycling, and drainage rates are balanced. Manipulation of any one of the mutually adjusted factors will lead to the alteration of the others. However, because of the different responses involved the effects will not be the same in each case. Furthermore, it must be reiterated that the sensitivity of a state of a system to alteration in external conditions will increase in proportion to the importance of the condition in maintaining that state. The picture also depends upon the degree of disturbance occasioned by any fluctuation.

Thus, if biological cycling is changed by, say, total vegetation removal, then the effect on nutrient status of the overall system will depend upon what proportion of the nutrients in the system are tied up in the biological cycling. If most of the nutrients in the system are retained in the vegetation and not in the soil it can be expected that upon vegetation removal a drastic loss of nutrients would occur in the system as a whole. Conversely, if most of the nutrients in the overall system are located in the soil subsystem, then release from exchangeable cations and weathering can occur and the losses can be offset; the losses will be temporary and the system can soon recover. It is true that the former system, where all the nutrients are in the vegetation, can recover, but the main source will be from rainfall and atmospheric contributions. Alternatively, vegetation may become gradually established and, especially if deep-rooted tree growth occurs, the rooting systems of the vegetation may be able to bring nutrients up from some depth in the soil and bring them into circulation. The different disturbance and recovery types are illustrated in Fig. 8.23. Figure 8.23(a) plots a system where the nutrients are held in the vegetation and Fig. 8.23(b) shows a system where the nutrients are held in the soil.

If the vegetation is altered only slightly, then a temporary set-back in cycling is liable to be offset by regrowth. Weathering releases and drainage losses are liable to be unaffected.

As a generalization, if the disturbance of the biotic component is *slight* then the *biotic* components are liable to be unaffected. If the disturbance of the biotic components is *large* or *total* then there is liable to be little, slow, or no recovery of the *biotic* components and the *abiotic* components are liable to be affected, both relatively to the biotic components and absolutely. The recovery, in the case of large or total disturbance, will depend more upon the reserves inherent in the abiotic components and the way in which the biotic components act upon them to mobilize them. It is like the pond in the original analogy (Section 8.2). If the pond represents the nutrient store and if this is

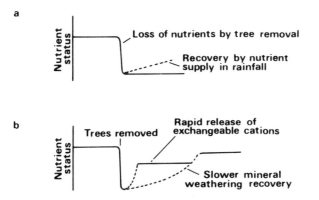

FIG. 8.23.   (a) Nutrients tied up in vegetation and not in soil (b) Nutrients in soil and not in vegetation

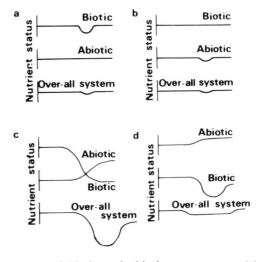

Fɪɢ. 8.24.   Disturbance of biotic and abiotic components. (a) Nutrients all in vegetation (b) Nutrients all in soil (c) Nutrients all in vegetation (d) Nutrients all in soil

slightly depleted the normal stream input will be able to offset this. If the pond is emptied or if the stream is stopped from flowing (or both) then it can be imagined that, for example, ground water may gradually seep into the pond to refill it. Thus a slower, normally less important, component becomes significant in the face of severe disturbance of the normal state of the system. Figure. 8.24(a) considers a slight disturbance where the nutrients are all in the vegetation; (b) considers a slight disturbance and the nutrients all in the soils; (c) and (d) consider large disturbances, (c) with the nutrients in the vegetation and (d) with the nutrients in the soil. The overall nutrient status of the total system is shown as well as the status of the biotic and abiotic components.

In general, much of the abiotic recovery will be effected by the chain of events pictured in Fig. 8.25. Assuming weatherable minerals to be present in the system, the system will eventually balance since it has both negative and positive feedbacks. Individual relationships are shown in Figs. 8.25(a), (b), and (c) and the overall interaction is shown in Fig. 8.25(d) where the + sign indicates a positive relationship between the variables and the – sign indicates a negative relationship.

The relationships of drainage loss to the weathering factors are shown in Fig. 8.26. Basically, drainage loss causes progressive nutrient release and eventual lowering of nutrient status since the system becomes unbalanced. There are two negative relationships shown in Fig. 8.26(a) and, since this is so, the two relationships shown in (b) and (c) can be expressed as in (d). Thus the overall system can be redrawn as in Fig. 8.26(e). The progression of

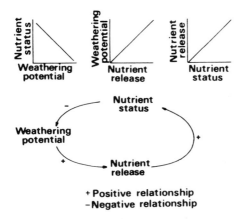

+ Positive relationship
− Negative relationship

Fig. 8.25.   Abiotic recovery—chain of events and feedback

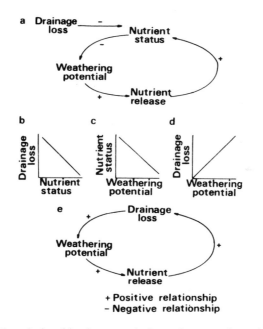

+ Positive relationship
− Negative relationship

Fig. 8.26.   The relationships between drainage losses and weathering factors

this positive feedback system will proceed until the limits of drainage rates are reached and output of nutrients will equal input or until the input is exhausted. In the former case the situation will be drainage rate limited and in the latter in will be input supply limited.

The effect of biological cycling on disturbances of the nutrient status caused by drainage disturbances will be to dampen the response patterns

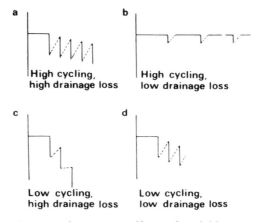

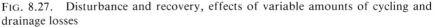

FIG. 8.27.   Disturbance and recovery, effects of variable amounts of cycling and drainage losses

already shown in Figs. 8.18 to 8.20, unless biological cycling is a small component relative to drainage losses. With large amounts of biological cycling it would simply mean that the vertical components of the diagrams in Fig. 8.18 to 8.20 would be lessened. As an adaptation of Fig. 8.20(a) the effects of variable amounts of cycling and drainage losses are shown in Fig. 8.27. In these diagrams (a) to (d) the downward line represents drainage losses of nutrients and the pecked, recovery, line is dependent upon nutrient cycling. Figure 8.27(b) represents the most stable system, (c) the least stable, and (a) is relatively stable because of the high recovery potential (but it will still show a progressive decline unless cycling and drainage are finely balanced).

Lastly, variations in rainfall amount will be discussed. Free drainage is assumed throughout. Simply, if rainfall is increased then the chain of events will be similar to that of Fig. 8.26 and the system will progress until the output of the system is adjusted to the rainfall input (or the weatherable

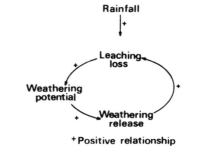

FIG. 8.28.   Rainfall effects, abiotic model

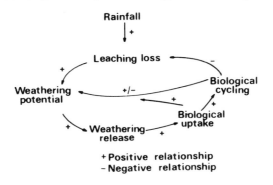

FIG. 8.29.   Rainfall effects, biotic model

minerals are exhausted), as shown in Fig. 8.28. Biological cycling will act to offset some of the leaching losses and stabilize the system at a higher level than would be apparent in an abiotic system (see Fig. 8.29—note that biological factors can influence weathering potential in either direction). According to the causal chain described in Fig. 8.29 the progression of the system through time will appear as in Fig. 8.30(a)–(c); (a) includes no cycling, (b) includes cycling, and (c) demonstrates the effects reduced nutrient status can have upon biologically induced weathering potential—stabilizing the system. It can be seen that biological factors will be crucial in stabilizing the nutrient status of the system.

8.5.4. Disturbance and recovery in nutrient systems—summary and conclusions

Two important points are these:

1. The sensitivity of a state of a system to fluctuations in external factors increases in proportion to the importance of those factors in maintaining that state.

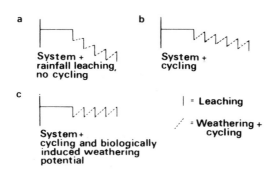

FIG. 8.30.   Rainfall effects, system over time

2. The above statement should be qualified by the fact that, under disturbance, systems may react in a way so as to alter the relative importance
of the factors in maintaining a given state. Primary maintaining
factors may decrease in importance (in which case the state may be
maintained at the same level by another factor operating in a similar
direction or a new, dominant factor may direct the system to a new
state). Alternatively, primary maintaining factors may increase in
importance in the face of the disturbance and thus reinforce the
system.

It is therefore important to evaluate:

1. The relative importance of external factors in maintaining any given
status of systems.
2. Which of these factors (and their effects) are being altered by disturbances of fluctuations in external conditions.
3. Whether or not the relative importance of the factors is being changed
by the disturbances.
4. If the relative importance is being altered whether the new dominant
factors are operating in the same or in opposite directions to the
original dominant factors.

If the dominance of factors remains the same then stressing of a dominant
factor will have far greater effect than the stressing of a subdominant factor.
If the dominance of factors alters under stress then the system could be
considerably reinforced or considerably alterd, depending upon the degree of
coincidence of the direction of the new dominant factor with the old.

The above statements will be true subject to the following qualifications
and variations:

1. The biotic and abiotic responses to disturbances are different, the
latter usually being slower.
2. The biotic components are more resilient in the face of slight disturbances than they are in the face of large disturbances. The abiotic
components are usually only changed by larger disturbances. These
differences are due to the differences in flows of energy and matter. A
biotic high-flow system is self-reinforcing in the face of relatively small
disturbances and can recover quite quickly; it cannot usually recover,
however, if the flow is damaged radically. Low-flow abiotic systems
are not so self-reinforcing in the face of disturbances but, because they
often have more passive reserves than high-flow systems (where there
are little reserves as all the matter and energy is tied up in flow), they
can use these reserves to effect a long-term recovery rather more
reliably than the high-flow systems. Under slight disturbance high-
flow systems thus show a better recovery than low-flow systems.
Under massive, drastic, or total disturbances low-flow systems may

show a better eventual recovery than high-flow systems, largely because the low-flow systems tend to have more passive stores in reserve than the high-flow systems.

The identification of active and passive components in the soil and vegetation systems is thus crucial. The active part may simply be the vegetation and the passive part the mineral store in the soil; or there may be various combinations of different subcomponents of the system, according to conditions. However, a generalization can be made: highly active involvement of total reserves facilitates active recovery in the face of subtotal disturbances but impairs recovery if the active part of the system is totally removed; low active involvement of total reserves and high passive storage impairs immediate recovery of the active part in the face of subtotal disturbance but facilitates long-term recovery of the active portion of the system in the face of massive or total disturbance of the active portion.

Herein lies the hub of the nature of stability of soil and vegetation systems. Three levels of disturbance can be envisaged:

(i) Disturbance of active parts, where short-term recovery is possible if active involvement is high. Long-term recovery is possible if active involvement is low and passive reserves are high.

(ii) Disturbance of reserves, where recovery is possible provided reserves are high.

(iii) Removal of reserves, where no recovery is possible.

For example, recovery of grass after trampling is possible if grass growth is high. If growth is low, recovery is impaired. If soil nutrient reserves are high, grass regrowth is encouraged. If, however, soil is removed, no recovery will be possible. Figure 8.31 illustrates this point in that with varying depths of system disturbance, decreasing possibilities of recovery will be seen.

3. Some nutrient systems are self-reinforcing and possess internal negative feedbacks (for example, as in Fig. 8.26). Others are not so and only become regulated over time by external factors such as drainage rates, rainfall rates, or the supply of weatherable minerals. Thus, on the one hand, there are systems which are dependent of weatherable minerals. Thus, on the one hand, there are systems which are dependent on external variables and will vary if they vary; on the other hand, there are systems possessing internal self-regulation and

FIG. 8.31. Variations in depth of disturbances and varying degrees of recovery

compensatory mechanisms which will not be so prone to variation in the face of external fluctuations. Most of the compensatory mechanisms involve biological growth and energy flow but some involve increase or decrease of chemical reactions and of weathering rates.

4. Some nutrient systems are over all more resilient to disturbance than others. This is because some are in stable equilibrium and some are in unstable equilibrium, the former usually being more resilient. The reason for resilience may be one of degree of input or of self-regulation. A system may be conditionally stable if it is only held in position by inputs—if the inputs vary, the equilibrium will alter. The system is more stable if it possesses internal self-regulatory mechanisms.

The higher the flow (or potential flow) of nutrients is in the system then the more resistant the system tends to be to changes in external factors which can alter nutrient status.

High flows of one factor mean that a system is extremely resistant to variations in factors other than the one involved in high flow but also that the system is more sensitive to variations in that, high-flow, factor than it is to variations in others. Fusing these points together, Fig. 8.32 shows a system where factor *A* is quantitatively important and factor *B* is quantitatively unimportant to the overall system; variations in *A* causes fluctuations in the overall system but not variations in *B*. Different degrees of reactivity of components are plotted in Fig. 8.33 and feedback and trigger mechanisms in Fig. 8.34.

The answer to the question 'Why are some systems more stable than others?' is not, then, a simple one. However, it is possible to specify the conditions under which the *flows of materials and energy are low relative to the disturbance*. In addition, the unstable systems are *those not possessing internal compensatory mechanisms capable of offsetting the effects of disturbances*. Simply, it is a matter of *action and reaction*. If reaction offsets the action, then the system is stable. Reaction depends upon the *rate of flow* and on *reserves of potential flow*. If rates of flow or reserves are low relative to the rates of the disturbances, then the system will be unstable. The system

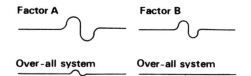

FIG. 8.32. Disturbance, factor A important, factor B unimportant, to over-all system

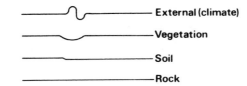

FIG. 8.33. Different degrees of reactivity of components

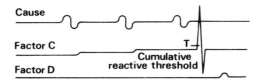

FIG. 8.34. Feedback and trigger mechanisms

will then change until the reaction is adjusted to the action, i.e. until the rates of flow are adjusted to the rates of disturbance or until an entirely new system obtains and different relationships and adjustments are seen.

The test of the value of the foregoing models is whether they can be used in management situations. Clearly, with reference to the work of Liddle (1975), discussed above, it is possible to identify unstable situations like a delicate arctic ground flora, with low rates of growth, which can be easily disturbed and altered under pressure. But away from these easily identifiable cases, it will be useful to cite a range of examples where the above generalizations and theories might be usefully applied in order to give a general basis to the understanding of the apparently varied natures of the problems involved.

8.6. Disturbance and recovery under anthropogenic pressures—examples

Some of the problems involved have already been touched upon by the postulation of such events as tree removal or vegetation modification. It will be useful in this section to elaborate upon these topics in the form of a brief discussion of several examples. The examples are listed below and will be dealt with in turn.

1. Footpath erosion.
2. Moorland reclamation.
3. Australian bush fires.
4. Deforestation.
5. Desertification in Africa.
6. Evolution of soils in Post-Glacial times.
7. Forestry plantations and soil acidification.

In each case-study the substantive details of the work involved are not given. Simply, an attempt has been made to recognize the type of disturbance and recovery pattern involved. For a fuller discussion reference can be made to some of the works involved, but the concern here is very much with theory, concept, and pattern rather than detail and mechanism.

8.6.1. Footpath erosion

In this case a progressive, repeated disturbance is often seen and two thresholds can be operative. One is where the vegetation is trampled beyond recovery and the second is where the soil is eroded. Final stability is thus defined when the system is exhausted and the soil has all gone. Conditional stability is present if the rate of trampling is less than or equal to the rate of plant growth. The over-all model is shown in Fig. 8.35.

Considering a series of equal trampling events, recovery is, at first, good. Productivity may even increase slightly (Liddle, 1975), but it begins to fall off with progressive trampling. Eventually a threshold is passed and the vegetation begins to disintegrate (this would equally well apply to other systems under pressure, like sand dunes). Although trampling becomes less diffuse, as people spread out over the path, the recovery potential drops towards zero as the system passes from biotic to abiotic and a rapid system change is seen. A second threshold is seen where all the soil has gone and a rock surface is present and then the system changes, but slowly over time.

Up to the time of the first threshold, recovery to the original state is possible by natural processes. Observations by Barkham (1972) suggest that different vegetation covers show different degrees of resilience and recovery. *Festuca* is thought to provide the most resilient cover, bracken (*Pteridium*) an intermediate cover, and *Sphagnum* has the least resilience. Thus the overall trajectory of a trampling recovery series would be as in Fig. 8.36, the speed of recovery and degree of resilience being related to vegetation type, as shown. Evaluation of thresholds for different species in relation to trampling intensity is thus of primary importance in the tackling of this problem (as, indeed, it is in other fields of recreational ecology).

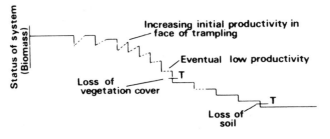

FIG. 8.35.   Biomass and trampling

FIG. 8.36. Disturbance and recovery of three vegetation types subjected to one trampling disturbance

8.6.2. Moorland reclamation

Work by Maltby (1975) and Maltby and Crabtree (1976) has shown an interesting pattern of land reclamation and reversion on Exmoor. The land was reclaimed around 1815 and some parts have been allowed to revert. This represents a disturbance and recovery sequence in the soil and vegetation system. The disturbances were represented by reclamation procedures (which are listed as paring-off the turf, burning, spreading of lime, ashes, and slag, ploughing, re-seeding, and the addition of fertilizers). The interest is in the identification of the varied time scales over which the varied factors have reverted. The data presented suggest the recovery pattern to be different for different components, as shown in Fig. 8.37, with the biotic components recovering before the abiotic. This is basically because soil type appears to be reverting subsequent to, and in response to, biological changes and in accordance with positive feedback mechanisms.

8.6.3. Australian bush fires

Work has been undertaken by Henderson and Wilkins (1975) and they suggest that there is an optimum time spacing of bush fires in the Australian region studied. Here, the fires act to regulate the amount of brushwood and litter which collects. Basically, the theory is that small, frequent fires act to maintain an equilibrium: if fires are prevented by man, brush and litter accumulate to such an extent that such fires as do occur are far more devastating than the small, frequent fires which occur naturally. The problem is that

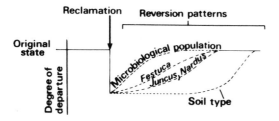

FIG. 8.37. Disturbance and recovery of vegetation and soil components of moorland during reclamation and reversion

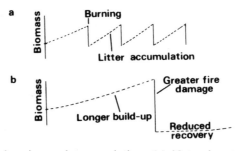

FIG. 8.38. Litter burning and accumulation. (a) Natural system (b) Under fire control scheme

the larger fires appear to damage some of the regeneration stock as they burn more fiercely rather than sweeping quickly and lightly through the bush. Thus the heavy burning delays the recovery of the system. As a management policy, a return to non-prevention of natural fires is advocated and this will allow the system to be self-maintaining. The system with and without man's disturbance is shown in Fig. 8.38. Fire and ecosystems are discussed in general terms by Kozlowski and Ahlgren (1974) and a British example is given in Imeson (1971).

8.6.4. Deforestation

Deforestation results in direct nutrient loss from the soil and vegetation system by biomass harvesting. In addition, because of the decay of cut material and the loss of system retention by recycling, solute output increases directly after deforestation but gradually decreases during regrowth (Swank, 1986). Losses in sediments also increase, but more gradually initially as protective, cut material rots, followed by erosion from bare surfaces, decreasing again as the plant cover is re-established. This is shown for the Hubbard Brook catchment. northeast USA, by Bormann *et al.* (1974), and an illustration of the progression through time of the nutrient status in the soil and vegetation system is shown in Fig. 8.39.

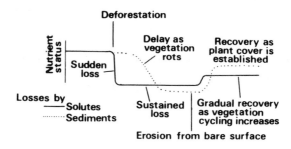

FIG. 8.39.   Disturbance and recovery of solute output Hubbard Brook

Table 8.1 *Nutrient losses from undisturbed and felled hardwood forest catchments*

Nutrient* (ppm)	Undisturbed Hardwood	Clearcut**	10-Year Old Coppice
NO$_3$−N			
mean	0.008	0.094	0.149
max	0.050	0.208	0.246
PO$_4$−P			
mean	0.086	0.004	0.006
max	0.029	0.020	0.095
K			
mean	0.589	0.456	0.426
max	1.368	1.051	0.975
Na			
mean	1.351	0.901	0.698
max	1.676	1.692	1.231
Ca			
mean	0.749	1.021	0.742
max	1.244	1.319	1.128
Mg			
mean	0.359	0.275	0.312
max	0.552	0.675	0.637

*data are for maximum means and maximums recorded from a number of catchments.

**Mostly clearcut, but includes some thinned and undisturbed areas.

Source: Douglass and Swank (1975)

More extensive work on the effect of deforestation on water and solute output has been undertaken in the Coweeta catchment, North Carolina, USA (Douglass and Swank, 1975; Swank, 1986). Over a 40 year monitoring period they showed that there were distinct increases in flow from the catchment, most noticeable at times of low flow and largely in relation to decreased interception losses and decreased evaporation losses after canopy removal. They also charted some increases in nutrient losses following cutting. Table 8.1. summarizes their data from a number of catchments, with the maximum figure quoted from their range of means and maximums.

Many solutes show no greater losses, or decreased loss, in the last two columns for clearcut and coppicing, specifically phosphate–phosphorus, potassium, sodium, and magnesium. Nitrate–nitrogen showed the most significant and sustained increased loss following disturbance. Calcium losses also increased after clearcutting but were at pre-felling levels in 10-year old coppice. Thus, the general diagram (Fig. 8.39) does not apply equally to all solutes but, as might be expected, applies more to nitrate–nitrogen which will be the most closely involved in biological processes, such as fixation and

Table 8.2 *Disturbance and recovery of water and nutrient outputs following clear-cutting and logging at Coweeta, eastern USA. Nutrients in kg ha^{-1}*

Year*	Flow (cm)	NO$_3$−N	K	Ca
1	2.8	0.03	0.84	1.85
2	26.5	0.26	1.98	2.60
3	20.5	1.12	1.97	2.53
4	17.3	1.27	2.41	3.17
5	11.9	0.25	0.88	1.66

*Data are for water years, May–April; the first year involves only 4 months of cutting.
Source: Swank (1986)

mineralization, and therefore will be liable to be more affected by vegetation disturbance.

Data provided by Swank (1986) suggest that at least a five-year period is necessary for recovery of solute outputs from logging and that as well as perturbations of nutrient output from losses of cycling by vegetation removal, decreased interception also increases runoff. The data are shown in Table 8.2. Flow maximum was in year 2 following disturbance and was 9.5 times higher than year 1; the nitrate–nitrogen peak was 42 times higher at the peak in year 4, potassium 2.9 times higher than year 1 in the peak at year 4, and calcium was 1.7 times higher, also in year 4. In year 5 flow was still 4.25 times higher than in year 1, nitrate was 8 times higher, potassium had recovered to only 1.05 times higher, and calcium was lower at 0.9 of year 1 level. Thus, flow appears to take longer to recover than nutrient export, and nitrogen losses appear to take longer to recover than potassium or calcium. These data can be represented in the form of a disturbance and recovery model for solute output as shown in Fig. 8.40.

8.6.5. Desertification in Africa

In the study by Rapp (1974) climatic change is the basic disturbance factor. The interest is in the successive hierarchy of environmental responses. Slight fluctuations in climate lead to slight variations in river and lake levels as well as plant productivity and grazing capacity. Cumulative effects are seen, however, in marginal situations. One of the important factors is that recovery after drought may be inhibited in marginal areas by the activities of man. Clearing, burning, and overgrazing markedly limit the capacity of vegetation to recover, even though post-drought conditions may revert to ones which are favourable to vegetation growth. The relationships between rainfall variations and number of cattle are illustrated in Fig. 8.42(a). A further interesting cycle is shown in Fig. 8.42(b). A summary diagram is presented in Fig. 8.42(c).

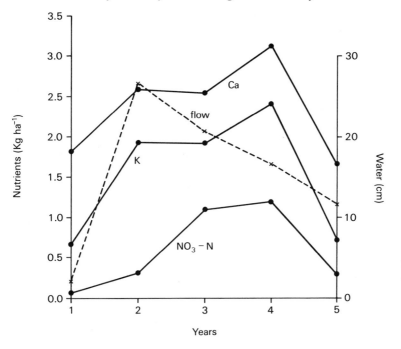

Fig. 8.40   Disturbance and recovery of water flow and nutrient export from a forested catchment, Coweeta, USA, after clearcutting and logging in year 1. Drawn from data from Swank (1986)

8.6.6. Evolution of soils in Post-Glacial times

This is a widely discussed topic and, amongst others, it is mentioned by Curtis *et al.* (1976), Dimbleby (1962), and Gimingham (1972). Basically, as far as can be generalized, it seems that during the Atlantic period increased rainfall input was leading to soil impoverishment (along the lines suggested by Figs. 8.28 and 8.29). The effect of man in forest clearance was to accelerate this process by the removal of vegetation, the reduction of the biological cycling component, and the reduction of nutrient retention by cycling. Thus a dual response is seen, firstly in the biotic component and secondly in the abiotic, as shown in Fig. 8.42. In more detail, the removal of vegetation by man would cause recovery from leaching to be lessened until a threshold was reached and the system graduated into a new soil type (Fig. 8.43).

8.6.7. Forestry plantations and soil acidification

It is often claimed that afforestation with conifers leads to soil acidification. In fact this depends on the tree species and the nature of the soil. Tree species

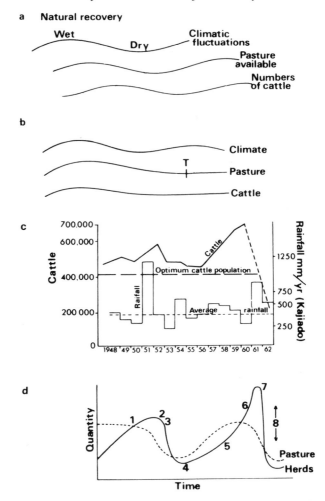

FIG. 8.41. Desertification. (a) Climatic change, natural vegetation recovery. (b) Climatic change, lack of recovery if vegetation regeneration is discouraged by grazing activity. (c) Cattle population and rainfall at Kajiado, Kenya, 1948–62. Years with above-average rainfall (1948, 1951) show an increase in cattle numbers due to abundant grass growth. Dry years cause a drop in cattle numbers. The increase in cattle numbers was extremely high in the years 1956–9 owing to rich grazing plus cattle vaccinations. The drought years of 1960–1 were disastrous and most of the 700,000 cattle died from starvation. The drought ended in November 1961 with extremely heavy rains which brought the annual figure up to well above average. (d) Generalized diagram of cattle numbers and pasture after disease control campaign. 1. Herds at carrying capacity. 2. Animals weakened by food shortage and susceptible to disease. 3. Disease outbreak; high mortality. 4. End of disease; all animals vaccinated to prevent further outbreaks. 5, Herds below carrying capacity grow quickly. 6. Herds greatly exceeding carrying capacity. 7. Pasture exhaustion: high mortality. 8. Same cycle as before the disease campaign but with wider fluctuations ((c) and (d) from Rapp, 1974)

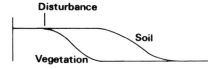

FIG. 8.42. Dual reaction, firstly biotic, secondly abiotic

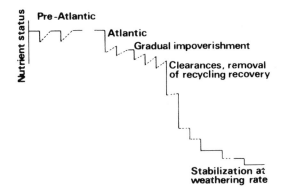

FIG. 8.43. Postulated Post-Glacial deterioration of soil

can vary markedly in their leaf acidity (as already shown in Fig. 4.3 and it is clear that there is an overlap between some deciduous trees and some conifers. Moreover, some conifers are more acidic than others. In addition, soils which are already acid will show very little additional effect and alkaline soils, with a large store of weatherable minerals, will also show very little effect. It is the slightly acid and circumneutral soils which will show the most affect (as already discussed for acid rain, Section 4.4.2). Nevertheless, it is on the soils and for the most acid-leaved conifers, that the acid input represents a disturbance to the system which can often not be offset. In many ways, the acidification acts through a decrease in earthworm population which would be important in turning the soil over and bringing back leached nutrients to the surface. As soil acidification increases, earthworm population decline, reinforcing the acidification trend (Fig. 8.44). Thus, the acidification of soils by conifers is of interest since it represents an increase in weathering potential owing to the acid nature of the litter. Thus nutrient status is lowered by weathering mobilization until the output stabilizes against the input (Fig. 8.45). Planting of other species may increase nutrient status as they may be deeper rooted than the pre-existing vegetation. Mineral reserves at depth may therefore be tapped. Thus planting a heathland with trees may reverse the low nutrient status (Fig. 8.46); conversely, tree felling and the establishment of a low nutrient status may occur under a shallow rooted vegetation. On sand dune soils, Wright (1955, 1956) charts an increase in leaching

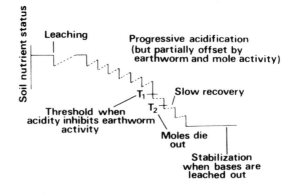

FIG. 8.44.   A generalized model of progressive soil acidification

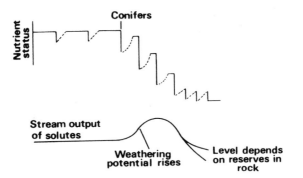

FIG. 8.45   Afforestation—soil property changes after coniferous afforestation

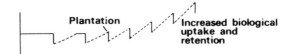

FIG. 8.46   Afforestation—soil changes after afforestation of heath

following stabilization, but an increase in nutrient status of the surface soil as deep rooting systems become established, pulling up nutrients from depth and redepositing them in litter on the soil surface.

8.7. Conclusions

Many different forms of stability and recovery are exhibited in soil and vegetation systems. Details have not been supplied but the examples are given in the conviction that they illustrate an *important way of thinking*. General themes have emerged. The rate of change of an external factor relative to the potential rates of reaction in the system is important. The differing rates of response of the various components of the system also emerge. Some case-studies have exhibited progressive changes and these pose special problems for management policies. If the desired state is to be achieved then the negative feedbacks in a system have first to be identified and then to be amenable to operation by man (see Chorley, 1973). In the case of progressive, positive feedback a negative feedback may have to be artificially induced to maintain the system at a desired level.

Stability and recovery ideas in soil and vegetation systems are basically simple in conception but complex in their variety. Stablity and instability may arise for many different reasons and in the face of many different types of disturbances. However, two things are important. One is that in the environmental sciences thought should be given to stability and disturbance along the lines of discussion suggested above. The second is that experimental work is needed on the quantitative importance of various factors involved in environmental processes. Prediction of system stability would be made less difficult if this type of knowledge was more readily available. While it is laudable to work on applied problems, like pressure of recreation upon footpaths, or sand dunes, or on agricultural ecosystems, these studies will be built on cardboard foundations if the basic, underlying principles are not understood. More, possibly seemingly unapplied, work will have to be undertaken on the relative importance of factors in their contributions to over-all system states, as well as on the more direct stability under pressure studies, before stability and recovery in soil and vegetation systems is better understood.

Further reading

Pickett and White, 1985.

9 Conclusions: a Sense of Problem

Necessity is something which exists in the mind, not in objects.

HUME

The original aim of this book, as stated in the Preface, was to attempt to study rock, soil, and vegetation together as an interactive unit. Yet only a small portion of rock–soil–vegetation interactions have been studied here because the amount of information available is so great that some selection is necessary in order to achieve some degree of comprehension. Thus, in many ways, the focus has been upon concepts and ways of thinking, rather than upon substantive information. Indeed, it would be extremely difficult for any one person, or even group of people, to try to achieve a total view at a detailed level of information. The problem is that, as discussed in Chapter 1, because of the inevitable selection of information, transdisciplinary statements, while desirable, tend to be superficial. On the other hand, specialized statements can carry greater depth and conviction. However, these tend to be valid only within carefully prescribed limits and thus are often of limited application. The problem can be illustrated by a diagram (Fig. 9.1). Knowledge can be thought of in terms of two axes—the horizontal axis representing breadth and interrelationship and the vertical axis representing depth and specialism. Thus a T-shaped diagram emerges. Horizontally, there may be a knowledge of, for example, soil type, plant type, and rock type. Here, cross-reference can readily be made without much specialized knowledge. At this level statements such as the following are possible: that *Sesleria* grass grows on calcareous soils which develop on limestone rock or that *Juncus effusus* tends to grow on poorly drained gley soils which tend to develop on clay strata. Within each subject (of soil science, botany, and geology) there is a depth of

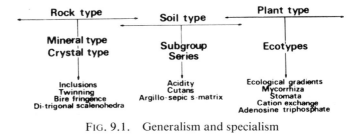

FIG. 9.1.   Generalism and specialism

specialized knowledge, with increasing detail with increasing depth. For example, there is knowledge of soil subgroups; of plant ecotypes, and of mineral type and crystal form. Below this level there is a knowledge of greater specialism—soil plasma types, plant root fungi, crystal birefringence, and so on.

In many cases it may be possible to relate two or more specialized pieces of knowledge—like root cation exchange and soil acidity—but this cross-referencing in itself is specialized and may remain isolated from the rest of environmental knowledge. What is necessary is not just the investigation of relationships between a few individual components of the environment but the relating of detailed knowledge across the board in an overall context. This is especially true if the broad, correlative statements such as those given above about plant type, soil type, and rock type are to be given any functional meaning, because correlation does not necessarily prove cause and effect. Detailed process knowledge from many subjects is necessary. Thus the horizontal T piece of the diagram needs to be brought downwards as far as possible. In other words, the need is to try to relate as much knowledge from one subject to as much knowledge as possible from all other subjects, at the most detailed level possible.

If this goal were to be achieved, then the researcher would be in a very powerful position to lay out valuable management policies. Clearly, however, this goal remains difficult to conceive of intellectually—let alone actually achieve, even with the aid of computers. Yet this should of course be one of the ultimate aims of the scientific investigation of the environment—to investigate and understand in detail how a mass of individual components interact.

There are two things which can be achieved at the present state of knowledge. These are first, the extension of linkages from component to component and second, the building-in of more detail to the more general models. While this book has attempted to do both, clearly the latter is probably the most feasible and the one that has been the least difficult here. Having attempted this kind of work on a conceptual basis it is now possible to suggest some requirements for future work.

1. Research should study the effect each component has upon (*a*) each other component in the system and (*b*) the overall system comprised of these components.
2. Research should identify the components which are important in controlling the over-all system and distinguish these from the less important ones.
3. It will be necessary to identify the external factors which influence the internal state of the system and it should be possible to quantify the degree to which these external factors affect the fundamental relationships and the degree to which they affect quantitative relationships (if

the actual nature of the relationship is not itself affected).
4. The researcher should be able to predict the future states of the system and especially the stability of the system in relation to external events.
5. Ultimately, the research worker should be in a position to identify how man's actions may affect relationships in a quantitative sense, or a fundamental sense, and to recommend viable management policies.

There are a number of problems which arise from these requirements. In many ways the main problem is one of focus (Fig. 9.2). With a limited focus, a research project becomes amenable to manipulation and rigorous specification, but a research project designed with a greater sense of awareness of interrelationship a certain loss of focus may be experienced. If the research problem is limited to the topic of rock weathering and if this is the focus of the study then the models derived from any research undertaken will be relatively simple ones where weathering can be viewed as the result of a number of factors (for example, soil type, climate and vegetation: Fig. 9.2(a)). A wider

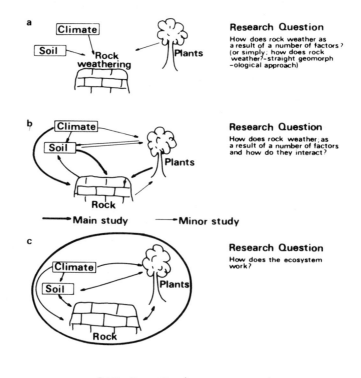

FIG. 9.2.   Approaches to environmental systems. (a) Limited research topic with a focus on one process (b) Research topic with one focus, but aware of interaction (c) More complex interactive environmental study (no particular focus)

state of awareness can be achieved (Fig. 9.2(b)) if the research worker has the knowledge that the perceived causative factors are interrelated. Here it is realized that, for instance, plant growth and climate are related and thus the weathering models produced will be more sensitive to interaction and the operation of multiple processes. These types of models are at a higher level of awareness. But if a greater awareness of environmental interaction is achieved (Fig. 9.2(c)) then the model gained is nearer a mutually reactive reality, but a certain loss of focus is experienced. Thus, if the aim of the research is clearly defined and limited as being, for example, about weathering, there is little problem of focus. But if the aim is the wider one of answering the question 'How does the environment work?' the research is more problematical. On one level it may become more diffuse and generalized, on the other it becomes subject to several different foci—i.e. each specific relationship is studied and the overall pattern is not necessarily seen. If a picture of the whole is to be built up then some filtering of information is necessary otherwise knowledge becomes too complicated and detailed to deal with easily. Thus the geomorphologist or ecologist has few problems in this context—each may even gain substantially from a knowledge of the other's discipline in tackling his own problem. The feeding-in of ecological information to a geomorphological problem, for example, can be extremely useful. The geomorphologist may even promote the understanding of the evolution of land-forms by studying overall ecosystems and environmental processes and then gain a knowledge of geomorphic processes as a spin-off. But he still has an ultimate geomorphological focus. However, the environmental scientist has a real problem. He is faced with researching into either broad, fairly obvious ecosystem models or immersing himself in a detailed congested network: he may thus end up either quantifying the obvious or hinting at the unobtainable. An answer to this apparent dilemma lies in the realization that, as human beings, we are limited by our faculties and that some degree of isolation, separation, or filtering is necessary in order to research into a topic (that is, within current paradigms of thought—and there may be others in the future where this may not present so much of a problem). What should not be forgotten, however, is that the isolated, separated, or filtered topic has a specific relationship with the real world and is not the real world itself. It is vitally important to feed the isolated and defined research work back into the total environment to establish whether the relationships described by delimited, controlled research are maintained in the overall system that exists in nature. The danger of not doing this is that our concepts of reality can become fossilized from the results of modelling isolated, experimental situations. The even greater danger is that management policies will be put into practice which are based on partial or isolated models. While this is inevitable to a certain degree, it should be borne in mind that the way we handle reality is based upon our concepts of reality. Thus there is a crucial point: our concepts should come not so much *from the models from the research topic we have isolated* in order to be able to handle it, but *from the way the model behaves when tested against reality*. Environmental research should therefore have three stages:

1. isolation and delimitation (in order to be able to deal with a problem);
2. careful scientific investigation and model production;
3. testing of the way the model behaves when re-inserted into the whole pattern.

Upon testing the following questions can be asked:

(a) Are there additional factors which mask the experimental relationships established?
(b) Are there additional factors which affect a component involved in overall interaction?
(c) Are there additional factors which fundamentally affect the relationship which was experimentally established?

The latter is obviously the most significant situation to identify. However simple this may seem in theory, this type of approach may be difficult to follow through substantially in practice. Moreover, there is still a residual feeling that analysis, testing, and synthesis fall short of discovering how a mutually interactive whole system behaves. At least a research worker, teacher, and student can go some way towards this discovery by going out into the field, not as a limited specialist but as one aware of the totality of nature. Therefore, a person, does not simply dig a soil pit and just look at the soil type in detail and dismiss the vegetation above as plants and as the things that botanists study; neither does one take quadrats of vegetation and discuss percentage dominance, genetic ecotypic variation and succession and dismiss soil as something soil scientists study; nor does one study runoff regimes of water, ignoring vegetation variations and soil type. One digs a hole in the soil, takes a quadrat and also studies water flow as component parts of an over-all system. The focus is on interrelationship between infiltration rates, percentage dominance of plant types and litter type and soil formation and plant type, etc. But more than this, the study is of mutual interaction. As has been discussed, it is difficult to discuss this without a theme or focus but the focus is not 'How does the soil form?' or 'How does the plant species composition vary?' nor even 'How do soil and plants vary together?' The question are those of function and of flow and of dynamics. Some of the most fundamental questions are 'What happens to the water in a system?', 'What happens to the nutrients?', and 'What happens in terms of energy flow?' It is a matter of looking at a field area and asking the following questions:

1. What is it like? (Measurement, observation, classification, co-variance study.)

2. Why is it like it is? (Measurement, experimentation, inference.)
3. How does it work? (Measurement, observation, experimentation, inference, modelling.)

It is important to ask the right questions and it may be better to have a broad answer to the important questions about the natural environment than a precise answer to a very limited topic of inquiry. The more precise answers can always be gained in the more rigorously defined experimental situations; less precise answers are often found in the mutually reactive environmental systems. Chemists can give precise answers to thermodynamic reactions between known minerals in defined chemical situations, but the step of testing whether their models still work in earth surface situations—where humus, bacteria, and growing plants are present—is not always made. Ecological botanists have perhaps gone furthest in their understanding of real-world interrelationships between mineral nutrition and plant distribution, especially by culture experiments (see, for example, Rorison, 1969). These experiments often include the presence or absence of specific minerals elements in order to elucidate the important factors in the overall situation. But, again, the interrelationships between leaf shape and size with the nature of the soil 'A' horizon and with infiltration capacity and with soil leaching and the chemistry of mineral solubilization are only partially known, if at all. It is the meso-scale knowledge which is lacking. Micro-scale evaluation of factors and broad-scale ecosystem modelling are possible but the middle ground of component interaction is substantially left untouched.

One approach to this middle ground is in the undertaking of field-work where factors are controlled by the judicial selection of sites. Field measurements of natural processes can be taken at two sites of, say, equal slope but with a variable bedrock or with the same soils but different slopes. Thus the importance of each factor to some other factor, like, say, runoff may be evaluated. However, this approach, while attractive, is often not possible, owing to the covariance of factors. Simply, because of mutual interaction many or all of the factors involved tend to vary together.

Thus the research worker is left with a sense of problem. The problems lie in the control of factors in analysis and in the difficulty of studying a mutually interactive system. It is to be hoped that in the future new techniques will be developed to tackle these problems. These may be statistical techniques or they may be related to technological developments in analysis. But the most important factor is to develop suitable ways of thinking and apposite points of view for dealing with environmental investigations. It is hoped that this book has stimulated the development of suitable points of view and increased awareness of some of the problems, even if it has not been able to solve them.

It is not the intention of the author to end by promoting one way of thinking or promoting study in any particular area or scale. Simply, it is felt that the research which will be profitable in the future is that which not only refers to both theory and practice and to both concepts and measurement techniques but also is that which can refer to all scales involved—the holistic macro-scale that can have detailed work built into it and the detailed work which makes reference to the whole are equally useful. There should, moreover, always be a dialogue between them in the mind of the research worker, teacher, and student in the environmental sciences. The study of mutually reactive wholes and the study of detailed controlled and isolated parts are equally rewarding but we cannot hope to progress in our under-standing of the environment without reference to both. It might be added that attitudes of mind are all important. Research workers should realize what their in-built assumptions are and, instead of carrying round their preconceptions and trying to force them into environmental situations, they should stand in the environment, in humility, and with an open mind (as far as is possible) and observe and see what they may learn. We cannot work without assumptions and preconceptions, but we should at least realize that they are there. We should also realize that necessity is something which exists in the mind, not in processes or objects. Obviously, this kind of conceptual and philosophical approach may be of little help to operational work in practice. However, it can be concluded that field-workers should both develop operational concepts which can be put into practice and realize the limitations of what they are doing. They should thus always bear in mind that while they are ostensibly finding something out, there may be something more to it than they think. Field-work and environmental investigation should always be a blend of feasibility, idealism, and theory. It is perhaps important not to let any of these dominate thinking too much or for too long.

Management decisions are always made in a state of partial knowledge and they may only be a matter of making guesses and going by them as long as they work well. However, management plans can be made more viable if they are based on sound field-work and on a sound body of environmental knowledge. This is especially so if the field-work and body of knowledge has a built-in awareness of complexity and of the limitations of the available knowledge.

With regard to current achievements and future prospects, the pointers to integrated study discussed in the first edition of this book (1977), such as those which could be found in the writings of Bormann and Likens (1969), Dale (1979), Johnson (1971), and Major (1970), have now borne fruit in the comprehensive analyses of ecosystems which have been written up and reviewed by several authors. They include Likens *et al.*, 1977, writing on the Hubbard Brook biogeochemistry study; Jordan (1985), writing on cycling in tropical forest ecosystems; Waring and Schlesinger (1985) on nutrient cycling; Swank (1986) on biological control of solutes in runoff; and also the reviews by Smith (1984) and Trudgill (1986). These writings have indeed, not highlighted the dilemma of the dichotomy between specialist details and broad models, but, as discussed in the Introduction to this book, they have

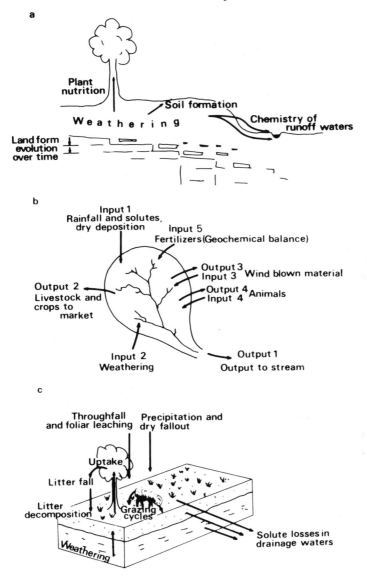

FIG. 9.3. (a) Weathering in an environmental context (b) Drainage basin approach to weathering (c) Geochemical circulation of whole ecosystems

shown that a fundamental unity can be seen in environmental processes. It is clear that such a holistic philosophy reduces the tension between specialization and generalization and these authors have shown how the specialist detail—at the bottom of the 'T' diagram (Fig. 9.1)—fits into the overall pattern of the system, as suggested in Fig. 9.2(c). The approaches which are important are summarized in Fig. 9.3. Fig. 9.3(a) shows weathering in an environmental context, 9.3(b) displays a drainage basin approach adopted by many workers, and 9.3(c) displays an approach to geochemical cycles in whole systems, including those affected by human activity. Thus, since the first edition of this book there have certainly been substantial developments in the type of research and outlooks promoted in this book. In this recent work, and it is to be hoped in future work, the attitude is one where circulations of nutrients, water, and energy are seen to operate together and hydrological processes and biological cycling provide the regulating mechanisms of the nutrient system of the soil and vegetation systems involved.

Stability, periodicity, and prediction are very important topics. In abiotic problems the issues have been discussed by Howard (1965), Schumm and Lichty (1965), Smith (1971), and Twidale *et al.* (1974). In biotic systems, Hurd *et al.* (1971) challenge the theory that greater diversity leads to greater stability. The edited works of Patten (1971, 1972, and 1975) together with the work of Dale (1970), and Pickett and White (1985), all have discussions on the use of systems analysis in simulation and prediction. Surveys like those of Duffey *et al.* (1974) continue to be useful (in the context of providing basic physical information), as do the management discussions with a physical basis (for example, Duffey and Watt, 1971) and also those with a broader basis (for example, O'Riordan, 1971). Mathematical simulation of soil-plant relationships are discussed by Kline (1973) and of leaching in soils by, for example, Frissel and Reiniger (1974). Thus advances are being made in concepts, application of systems analysis, surveys, predictions, and simulations. However, the basic need remains for further field measurements in integrated process studies. But as Slobodkin (1962) emphasizes: 'No single measurement is intrinsically significant. All measurements derive their interest from their context and the richness of predictive generalizations that can be produced from them.' In many ways the pioneering studies of the Hubbard Brook ecosystem, referred to several times in this book (Likens *et al.* 1977) have fulfilled many of the requirements of an integrated environmental study. Here, plant productivity, litter decomposition, stream solute levels, particulate losses, rain-water chemistry, chemical weathering, nutrient cycling, nutrient budgets, phytosociology, synecology, recovery, and disturbance are all among the topics which have been covered. This type of approach, with groups of research workers from different disciplines working together in one area, is extremely valuable. Overall ecosystem models can be worked out and specialized research undertaken by the investigation of a component of the overall system. Thus detailed work of limited scope can be fitted into the overall scheme.

The problem with this kind of approach is that a large outlay is necessary in terms of time, personnel, and equipment. Similar work, though of a less ambitious scale, has been undertaken by Crisp (1966) and Tallis (1964, 1965). However, this book will end with a consideration of field approaches which might be adopted by a student of soil and vegetation systems. Basically, the situation envisaged is that an investigator who is cognisant with the discussion in this book, goes out into the field and wishes to interpret what can be seen in a soil and vegetation system. The questions which can be usefully asked at a field site, at a number of levels of detail, are as follows:

1. *What happens to water on the site?* What is the rainfall in the area? How much rain-water falls directly on the soil and how much is intercepted by vegetation and stemflow? Is it liable to run overland, percolate into the soil, re-evaporate, be taken up by plants, percolate into groundwater, become stored in the soil, or run into a stream? What proportions of water will follow each of these routes and how much water flows in each soil horizon? How far will the picture alter with different weather conditions? How far are the runoff characteristics a function of static features of the site and how far are they a function of temporary features, like the storage of antecedent precipitation? What speed will the water travel at in different runoff routes? Given answers to the above (and they can be speculative or based on detailed measurements), what will the solute load of the runoff waters be and how does this vary over time?

2. *What is happening to the nutrients?* Does the soil parent material (unconsolidated material or bedrock) have a store of weatherable minerals? What chemical composition do the minerals have and how soluble are they in their environment? What is the mineral nutrient distribution within the soil profile? What is the mineral nutrient supply in the rain-water? What nutrients, and how much of each, are being lost in drainage waters? Which nutrients, and how many of each, are being taken up by plants? Are the nutrients in plants being stored in the vegetation, being leached from leaves down to the soil, being removed as a standing crop, returning to the soil in leaf litter, being leached and lost from the litter, or being taken up again by plants? Do organic matter and leaching act to encourage mineral breakdown? What is the status of available nutrients in the soil?

3. *What is happening to the organic matter?* How much primary production is there? How much litter fall is there? Is organic decomposition efficient—i.e. how much accumulation of undecomposed organic matter is there on the soil surface and in the soil profile? Are the soil biota present in large numbers and are they efficient in aiding organic cycling? How does biological activity vary with temperature and moisture?

These are only some of the main questions which can be asked and there are many more detailed ones which could be asked. They are divisible into the investigation of (1) static site characteristics and (2) temporarily variable characteristics. Furthermore, they are again divisible into (*a*) characteristics which are readily observed and easily measurable and (*b*) characteristics which can only be measured with difficulty or over long periods of research. Measurements and observations are clearly possible on a number of levels, but, subsequent to the definition of the system, the key questions are 'How does the whole system work?' and 'How stable is it?' Factors such as geology, soil type, and vegetation type can be observed directly. Secondary factors, like water route and residence time, solute load, nutrient routes, organic breakdown and soil biotic activity, while they could be investigated and measured, in lengthy research programmes, can also be estimated by deduction from observations. At any one time, measurements can be taken of soil pH, the nutrient content of soil, water, and plant material, and the soil content of organic matter and biota (see, for example, Allen, 1974; Jackson

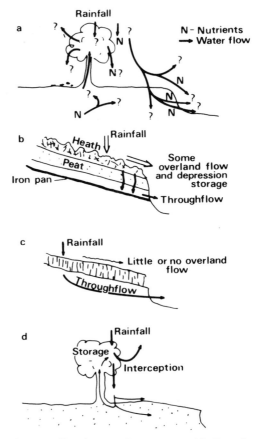

FIG. 9.4.   Approaches to soil and vegetation systems. (a) Questions to be asked of a soil and vegetation system (b) Some sample answers, showing relative proportions of water flow in upland heaths (the width of the arrow increasing with increasing proportion of flow) (c) Lowland grassland (d) Lowland woodland

and Raw, 1966). Other factors, such as runoff variation (Gregory and Walling, 1974) and variations in biological activity and solutes can only be measured over time. In teaching it is often sufficient to demonstrate that relationships and levels of concentrations exist, though making reference to the limitations of the representativeness of the data in a temporal sense. In order to prove any relationship with great rigour, more intensive sampling would have to be carried out, both in time and in space, usually with the statistical analysis of the data gained. Thus three different levels of approach can be identified: (i) simple thought processes and speculation about processes by field observation, deduction, and discussion; (ii) simple measurements which will illustrate the discussions, but which will have little or no statistical validity; (iii) detailed sampling programmes of the factors involved—either selecting a number of factors (because of time, financial, or personnel constraints) or as many factors as possible on an ambitious scale, such as is seen in the Hubbard Brook project. Some simple examples, at level (ii), are given in Fig. 9.4.

In summary, problems arise from the integrated study of soil and vegetation systems in that valid working models often have to be very complicated. However, integrated work points to an underlying unity in natural systems in that hydrological, nutrient, and organic systems often tend to operate together. As such, knowledge of one system can be used to promote the understanding of another. Field measurements of integrated processes are extremely useful and field-workers should bear in mind concepts and theory, as well as feasibility, so that units of field-work may be slotted in as components of overall models of soil and vegetation systems.

Glossary

Acid	A compound which yields hydrogen ions when it dissociates in solution, e.g. $H_2SO_4 = H_2^+ + SO_4^-$
Alkali	A compound which yields hydroxyl ions (OH^-) in solution.
Anion	Ion with a negative charge.
Base	A proton (H^+) acceptor, such as CO_3^{2-}, forming HCO_3^-.
Cation	Ion with a positive charge.
Chelation	Direct incorporation of a metal ion into the molecular structure of a compound, often organic (see Atkinson and Wright, 1967).
Cheluviation	Removal of metals in chelate compounds in gravitational water.
Congruent solution	The ions present in solution are in the same ratio as in the dissolving solid. Also incongruent solution, vice versa.
Dissociation	The separation of a compound in water into its component ions.
EDTA	Ethylenediaminetetraacetic acid—a powerful chelating agent.
Electrolyte	A weak electrolyte ionizes (dissociates) only to a slight extent in water; a strong electrolyte is almost completely dissociated.
Hydration	Incorporation of water molecules into crystal structure of a mineral, tending to cause its breakdown.
Hydrolysis	In chemistry: the reaction between ions of a salt and ions of water forming a solution which is either acid or alkaline. In the weathering of silicate minerals: this takes the form of the replacement of various cations by hydrogen ions from the solution, making the solution more alkaline (hydrogen ions are adsorbed into the mineral and it becomes unstable).
Ion	An atom which has acquired an electrical charge by the gain of an electron, making it negative; or the loss of a negative electron, making the ion positive (see also *Radical*).
Ion activity product	(IAP) Product of the activity (a measure of concentration) of the ions in solution.
Metal	A metal dissociates in water to yield positive cations, Ca^{2+}, Mg^{2+}, Al^3, Fe^{3+}, Na^+, etc.
pH	The negative logarithm of the hydrogen ion concentration in moles l^{-1}, where a mole is the molecular weight in grammes.
Radical	A compound behaving on ionization as if it were a single element, e.g. the compound SO_4^- behaving as one ion.
Redox potential	Oxidation—gain in oxygen (loss of hydrogen); loss of a negative electron giving a positive reaction. Redox Eh: $0.0—+0.8$. Reduction—loss of oxygen (gain of hydrogen); gain of a negative electron giving a negative reaction. Eh: $0.0——1.0$.

Salt	A salt is a compound which ionizes in water but which produces neither hydrogen nor hydroxyl ions in solution.
Sesquioxide	The ratio of the oxide to the metal is $1\frac{1}{2}$ to 1, e.g. Fe_2O_3 and Al_2O_3.
Solubility product	(K) The concentration of a compound in solution at equilibrium. The equilibrium is dynamic, not static, and solution = precipitation; (solution does not cease at equilibrium).

Abbreviations and units

$Ca^{2+} = Ca^{++}$ $\qquad\qquad$ $Al^{3+} = Al^{+++}$

p.p.m. = parts per million = mg l^{-1}

$^0/_{00}$ = parts per thousand

μm = micrometre or one-millionth of a metre or 0.001 mm (10^{-3} mm)

μmol = micro-mole

$\mu mol\ l^{-1}$ = milli mole per litre

μg = microgram

1 g = 1000 mg, 1 mg = 1000 μg, 1 p.p.m. = 1 mg l^{-1} = 1 mg kg^{-1}

1 ångstrom unit (Å) = 1×10^{-8} cm

0.0001 cm = 10^{-5} cm

600 000 cm = 6×10^5 cm

a = year

equivalent = molecular weight/valency

ha = hectare

Quick conversion chart for concentrations

$^0/_0$	$^0/_{00}$	$^0/_{000}$	$^0/_{0000}$	*p.p.m.*
1	10	100	1000	10 000

Normal solutions

1 mole l^{-1} = 1N

10 moles l^{-1} = 10N

1 mole = 1 gram mole = molecular weight in grams

(Refer also to Courtney and Trudgill, 1976 for an elementary introduction to soil science.)

Bibliography

ABATUROV, B. D., 1972, 'The role of burrowing animals in the transport of mineral substances in the soil', *Pedobiologia,* **12,** 261–6.

ACQUAYE, D., and Tinsley, J. 1965, 'Soluble silica in soils', in Hallsworth, E. G., and Crawford, D. V., *Experimental Pedology* (Butterworth), 126–48.

ADKINS, C. J., 1968, *Equilibrium thermodynamics* (Open University, McGraw-Hill).

ALEKIN, O. E., and Brazhnikova, L. U., 1968, 'Dissolved matter discharge and mechanical and chemical erosion', *I. U. G. G. —I. A. S. H. Symposium* (Bern, 1967), 35–41.

ALLEN, S. E. (ed.), 1974, *Chemical analysis of ecological material* (Blackwell).

—— Carlisle, A., White, E. J., and Evans, C. C., 1968, 'The plant nutrient content of rain water', *Journal of Ecology,* **56,** 497–504.

ANDERSON, J. L., and Bouma, J., 1973, 'Relationships between hydraulic conductivity and morphometric data of an argillic horizon', *Proceedings of the Soil Science Society of America,* **37,** 408–13.

ATKINSON, H. J., and Wright, J. R., 1967, 'Chelation and the vertical movement of soil constituents', in Drew, J. V. (ed.), *Selected papers in soil formation and classification* (Soil Science Society of America), 326–35.

ATTIWELL, P. M., 1966, 'The chemical composition of rainwater in relation to cycling of nutrients in mature Eucalyptus forest', *Plant and Soil,* **24,** 390–406.

BAAS-BECKING, L. G. M., Kaplan, I. R., and Moore, D., 1960, 'Limits of the natural environment in terms of pH and oxidation-reduction potentials', *Journal of Geology,* **68,** 243–84.

BACHE, B. W., 1983, 'The implications of rock weathering for acid neutralization', in *Ecological Effects of Acid Deposition,* National Swedish Environment Protection Board, Report, PM 1636, 175–87.

—— 1984, 'Soil–water interactions', *Philosophical Transactions of the Royal Society of London,* **B305,** 393–407.

—— 1985, 'Soil acidification and aluminium mobility', *Soil Use and Management,* **1,** 10–14.

BACK, W., and Hanshaw, B. B., 1965, 'Chemical geohydrology', in Ven te Chow (ed.), *Advances in hydroscience,* **2,** 50–109.

BARKHAM, J. P., 1972, 'Recreation and environment in part of the Lake District National Park', *Fieldworker,* Dec. 1971.

BEAR, F. E. (ed.), 1964, *Chemistry of the soil* (American Chemical Society, Monograph 16, Van Nostrand-Reinhold).

BECKWITH, R. S., 1955, 'Metal complexes in soils', *Australian Journal of Agricultural Research,* **6,** 685–98.

BEISHON, J., and Peters, G., 1972, *Systems behaviour* (Open University, Harper and Row).

BERTALANFFEY, L. VON, 1962, 'General systems theory: A critical review', *General Systems,* **7,** 1–20.

BERNER, R. A., 1971, *Principles of chemical sedimentology* (McGraw-Hill).

BERTHELIN, J., and Dommergues, Y., 1972, 'Effect of microbial metabolic products on the solubilisation of minerals in a granite sand', *Revue d'écologie et biologie du sol,* **9,** 387–406, from *Soils and Fertiliser Abstracts,* **36** (2), 1973, Abstract 678, p. 74.

BEVEN, K., and Germann, P., 1982, 'Macropores and water flow in soils', *Water Resources Research,* **18,** 1311–25.

BISQUE, R. E., 1961, 'Analysis of carbonate rocks for calcium, magnesium, iron and aluminium with E.D.T.A.', *Journal of Sedimentary Petrology,* **31,** 113–22.

BLACK, A. P., and Christman, R. F., 1963a, 'Chemical characteristics of fulvic acids', *Journal of the American Water Works Association,* **55,** 897.

—— 1963b, 'Characteristics of coloured surface water', *Journal of the American Water Works Association,* **55,** 753.

BLACK, C. A., 1968, *Soil-plant relationships* (Wiley).

BLOOMFIELD, C., 1950, 'Some observations on gleying: *Journal of Soil Science,* **1,** 205–11.

—— 1951, 'Experiments on the mechanisms of gleying', *Journal of Soil Science,* **2,** 196–211.

—— 1952, 'The distribution of iron and aluminium oxides in gley soils', *Journal of Soil Science, ,* 167–71.

—— 1953a, 'A study of podsolisation'. I: 'The mobilisation of iron and aluminium by scots pine needles', *Journal of Soil Science,* **4,** 5–16.

—— 1953b, 'A study of podsolisation'. II: 'The mobilisation of iron and aluminium by the leaves and bark of *Agathis australis* (Kauri)', *Journal of Soil Science,* **4,** 17–23.

—— 1954a, 'A study of podsolisation'. III: 'The mobilisation of iron and aluminium by Rimu (*Dacrydium cupressinum*), *Journal of Soil Science,* **5,** 39–45.

—— 1954b, 'A study of podsolisation'. IV: 'The mobilisation of iron and aluminium by picked and fallen larch needles', *Journal of Soil Science,* **5,** 46–9.

—— 1954c, 'A study of podsolisation'. V: 'The mobilisation of iron and aluminium by aspen and ash leaves; *Journal of Soil Science,* **5,** 59.

—— 1963, *Mobilisation and immobilisation phenomena in soils* (Rothamsted Experimental Station, Publication no. S10).

BLOW, F. E., 1955, 'Quantity and hydrologic characteristics of litter under upland oak forests in eastern Tennessee', *Journal of Forestry,* **53,** 190–5.

BOAST, C. W., 1973, 'Modelling the movement of chemicals in soils by water', *Soil Science,* **115,** 224–9.

BOCOCK, K. L., Gilbert, O., Capstick, C. K., Twinn, D. C., Wald, J. S., and Woodman, M. J., 1960, 'Changes in leaf litter when placed on the surface of soil with contrasting humus types'. I: 'Losses in dry weight of oak and ash leaf litter', *Journal of Soil Science,* **11,** 1–9.

BOHN, H. L., McNeal, B. L., and O'Connor, G. A., 1979, *Soil chemistry* (Wiley-Interscience).

BORMANN, F. H., and Likens, G. E., 1967, 'Nutrient cycling', *Science,* **155,** 424–9.

—— 1969, 'The watershed-ecosystem concept and studies of nutrient cycles', in Van Dyne, G. M. (ed.), *The ecosystem concept in natural resource management* (Academic Press).

—— 1970, 'The nutrient cycle of an ecosystem', *Scientific American,* **223,** 92–101.

—— and Eaton, J. S., 1969, 'Biotic regulation of particulate and solution losses from a forest ecosystem', *BioScience,* **19,** 600–10.

—— Siccama, T. G., Likens, G. E., Pierce, R. S., and Eaton, J. S., 1974, 'The export of nutrients and recovery of stable conditions following deforestation at Hubbard Brook', *Ecological Monographs,* **44,** 255–77. Copyright by The Ecological Society of America.

BORNKAMM, R., and Bennert, W., 1970, 'Chemical composition of the field layer—preliminary report', in Ellerby, H. (ed.), *Integrated and experimental ecology* (Ecological Studies, 2, Chapman and Hall), 57–60.

BOUMA, J., and Anderson, J. L., 1973, 'Relationships between soil structure characteristics and hydraulic conductivity', in Bruce, R. R., (ed.), *Field soil moisture regime,* Soil Science Society of America, Special Publication, No. 5, (American Society of Agronomy, Madison, Wis.), 77–105.

—— Jongerius, A., and Schoonderbeek, D., 1979, 'Calculation of saturated hydraulic conductivity of some pedal clay soils using micro-morphometric data', *Soil Science Society of America, Journal,* **43,** 261–4.

—— Dekker, L. W., and Muilwijk, C. J., 1981, 'A field method for measuring short-circuiting in clay soils', *Journal of Hydrology,* **52,** 347–54.

BOURGEOIS, W., and Lavkulich, L. M., 1972, 'A study of forest soils and leachates on sloping topography using tension lysimeters', *Canadian Journal of Soil Science,* **52,** 375–91.

BOYLE, J. R., and Voigt, G. K. 1973, 'Biological weathering of silicate minerals', *Plant and Soil,* **38,** 191–201.

BOYNTON, D., and Compton, O. C., 1944, 'Normal seasonal changes of oxygen and carbon dioxide percentages in gas from the larger pores of three orchard soils', *Soil Science,* **57,** 107–17.

BRADY, N. C., 1974, *The nature and properties of soils* (Macmillan).

BRAY, L. G., 1975, 'Recent chemical work in the Ogof Ffynnon Ddu system: further oxidation studies', *Trans. of the British Cave Research Association,* **2,** 127–32.

BREGER, I. A. (ed.), 1963, *Organic geochemistry* (Pergamon).

BREMER, J. M., Heintree, S. G., Mann, P. J. G., and Lees, M., 1946, 'Metallo-organic complexes in soils', *Nature* (London), **158** (4022), 790–1.

BRICKER, O. P., and Garrels, R. M., 1967, 'Mineralogical factors in natural water equilibria', in Faust, D., and Hunter, J. V., *Principles and applications of water chemistry* (Wiley).

BROECKER, W. S., and Oversby, V. M., 1971, *Chemical equilibria in the earth* (McGraw-Hill).

BUCKMAN, H. O., and Brady, N. C., 1969, *The nature and properties of soils* (Macmillan; 7th edn.).

BULLOCK, P., 1971, 'The soils of the Malham Tarn area', *Field Studies,* **3** 381–408.

BURD, J. S., 1925, 'Relation of biological process of cation concentration in soils', *Soil Science,* **20,** 269–83.

BURT, T. P., and Trudgill, S. T., 1985, 'Soil properties, slope hydrology and spatial patterns of chemical denudation', ch. 1 in Richards, K. S., Arnett, R. R., and Ellis, S., *Geomorphology and soils* (George Allen & Unwin).

CALVER, A., Kirkby, M. J., and Weyman, D. R., 1972, 'Modelling hillslope and channel flows', in Chorley, R. J. (ed.), *Spatial analysis in geomorphology* (Methuen), 197–218.

CALVIN, M., and Martell, A. E., 1953, *Chemistry of the metal chelate compounds* (Prentice-Hall).

CARLISLE, A. H. F., and White, E. J., 1966, 'The organic matter and nutrient elements in the precipitation beneath a sessile oak (*Quercus petraea*) canopy', *Journal of Ecology,* **54,** 87–98.

CARROLL, D., 1962, 'Rainwater as a chemical agent of geologic processes—a review', *United States Geological Survey, Water Supply Paper, 1535—G.*

CARSON, M., and Kirkby, M. J., 1972, *Hillslope form and process* (Cambridge University Press).

CHABEREK, S., and Martel, A. E., 1959, *Organic sequestering agents* (Wiley).

CHAPMAN, S. B., 1967, 'Nutrient budgets for a dry heath ecosystem in the south of England', *Journal of Ecology,* **55,** 677–89.

CHAPPELL, H. G., Ainsworth, J. F., Cameron, R. A. D., and Redfern, M., 1971, 'The effect of trampling on a chalk grass ecosystem', *Journal of Applied Ecology,* **8,** 869–82.

CHILDS, E. C., 1969, *The physical basis of soil water phenomena* (Wiley).

CHORLEY, R. J., 1962, 'Geomorphology and general systems theory', *United States Geology Survey Professional Paper,* **500-B,** 1–10.

—— 1969, 'The role of water in rock disintegration', in Chorley, R. J. (ed.), *Water, Earth and Man* (Methuen), 3.ii.

—— 1973, 'Comments on "Systems modelling and analysis in resource management" by J. N. R. Jeffers', *Journal of Environmental Management,* **1,** 29–31.

—— and Kennedy, B. A., 1971, *Physical geography: A systems approach* (Prentice-Hall).

CLARKSON, D. T., 1974, *Ion transport and cell structure in plants* (McGraw-Hill).

CLEAVES, E. T., Godfrey, A. E., and Bricker, O. P., 1970, 'Geochemical balance of a small watershed and its geomorphic implications', *Bulletin of the Geological Society of America,* **81,** 3015–32.

COETZEE, J. F., and Ritchie, C. D. (eds.), 1969, *Solute–solvent interactions* (Marcel Dekker).

COLE, D. W., 1968, 'A system for measuring conductivity, acidity and rate of flow in a forest soil', *Water Resources Research,* **4,** 1127–36.

—— Gessel, S. P., and Held, E. E., 1961, 'Tenison lysimeter studies of ion and moisture movement in glacial till and coral atoll soils', *Proceedings of the Soil Science Society of America,* **25,** 321–5.

COMMONER, B., 1968, 'The balance of nature', *Journal of the Soil Association,* 7.

COOKE, G. W., 1982, *Fertilizing for maximum yield* (Granada).

COSBY, B. J., Hornberger, G. M., and Galloway, J. N., 1985a, 'Modelling the effects of acid deposition: Assessment of a lumped parameter model of soil water and streamwater chemistry', *Water Resources Research,* **21,** 51–63.

—— —— —— and Wright, R. F., 1985b, 'Time scales of catchment acidification', *Environmental Science and Technology,* **19,** 1144–9.

COULSON, C. B., Davies, R. I., and Lewis, D. A., 1960a, 'Polyphenols in plant humus and soil'. I: 'Polyphenols of leaves, litter and superficial humus from mull and mor sites', *Journal of Soil Science,* **11,** 20–9.

—— 1960b, 'Polyphenols in plant humus and soil', II: 'Reductive transport by polyphenols of iron in model soil columns', *Journal of Soil Science,* **11,** 30–44.

COURTNEY, F. M., and Trudgill, S. T., 1976, *The soil: An introduction to soil study in Britain* (Edward Arnold).

CRABTREE, R. W. and Trudgill, S. T., 1981, 'The use of ion-exchange resin in monitoring the calcium, magnesium, sodium and potassium contents of rainwater', *Journal of Hydrology,* **53,** 361–5.

CRICKMAY, C. H., 1974, *The work of the river* (Macmillan).

CRISP, D. T., 1966, 'Input and output of minerals for an area of Pennine moorland: The importance of precipitation, drainage, peat erosion and animals', *Journal of Applied Ecology,* **3,** 314–27.

CROMPTON, E., 1960. 'The significance of the weathering/leaching ratio in the differentiation of major soil groups with particular reference to some very leached brown earths on the hills of Britain', *Report of the 7th International Congress of Soil Science,* **IV,** 406–12.

CRONAN, C. S. and Schofield, C. L., 1979, 'Aluminium leaching response to acid precipitation: effects on high-elevation watersheds in the Northeast', *Science,* **204,** 304–6.

CRYER, R., 1976, 'The significance and variation of atmospheric nutrient inputs in a small catchment system', *Journal of Hydrology,* **29,** 121–37.

CRYER, R., 1986, 'Atmospheric solute inputs', in Trudgill, S. T., (ed.), *Solute processes* (Wiley), 15–84.

CURTIS, C. D., 1976a, 'Stability of minerals in surface weathering reactions', *Earth Surface Processes,* **1,** 63–70.

—— 1976b, 'Chemistry of rock weathering: Fundamental reactions and controls', in Derbyshire E. (ed.), *Geomorphology and Climate* (Wiley), ch. 2.

CURTIS, L. F., and Trudgill, S. T., 1974, 'The measurement of soil moisture', *British Geomorphological Research Group, Technical Bulletin,* **13.**

—— Courtney, F. M., and Trudgill, S. T., 1976, *Soils in the British Isles* (Longman).

DALE, M. B., 1970, 'Systems analysis and ecology', *Ecology,* **51,** 2–16.

DAMMAN, A. W. H., 1971, 'Effect of vegetation changes on the fertility of a Newfoundland forest site', *Ecological Monographs,* **41,** 253–70.

DAVIES, R. I., 1971, 'Relation of polyphenols to decomposition of organic matter and to pedogentic processes', *Soil Science,* **111,** 80–5.

DAVIS, S. N., 1964, 'Silica in streams and groundwater', *American Journal of Science,* **262,** 870–91.

DEEVEY, E. S., 1970, 'Mineral cycles', *Scientific American,* **223,** 148–58.

DEJU, R. A. 1971, 'A model of chemical weathering of silicate minerals, *Geological Society of America, Bulletin,* **82,** 1055–62.

—— and Bhappu, R. B., 1965, 'Surface properties of silicate minerals', *New Mexico Institute of Mining and Technology, State Bureau of Mines and Mineral Resources, Circular,* **82,** 67–70.

—— 1966, 'A chemical interpretation of surface phenomena in silicate minerals', *New Mexico Institute of Mining and Technology, State Bureau of Mines and Mineral Resources, Circular,* **89,** 1–13.

DICKINSON, C. H., and Pugh, G. J. F. (eds.), 1974, *Biology of plant litter decomposition,* vols. 1 and 2 (Academic Press).

DIMBLEBY, G. W., 1952, 'The historical status of moorland in north-east Yorkshire', *New Phytologist,* **51,** 349–54.

—— 1962, *The development of British heathlands and their soils* (Oxford Forestry Memoirs, Clarendon Press).

—— and Gill, J. M., 1955, 'The occurrence of podzols under deciduous woodland in the New Forest', *Forestry,* **38,** 95–106.

DOMENICO, P. A., 1972, *Concepts and models in groundwater hydrology* (McGraw-Hill).

DOUGLAS, I. 1968, 'Field methods of water hardness determination', *British Geomorphological Research Group, Technical Bulletin,* **1.**

—— 1972, 'The geographical interpretation of river water quality', *Progress in Geography,* **4,** 1–81.

DOUGLASS, J. E., and Swank W. T., 1975, 'Effects of management practices on water quality and quantity: Coweeta Hydrological laboratory, North Carolina', *Northeastern Forest Experiment Station, USDA Forest Service General Technical Report,* NE-13, 1–13.

DUFF, R. B., Webley, D. M., and Scott, R. O., 1963, 'Solubilisation of minerals and related minerals by 2-ketogluconic acid-producing bacteria', *Soil Science,* **95,** 105–14.

DUFFEY, E., and Watt A. S., (eds.), 1971, *The scientific management of animal and plant communities for conservation* (Blackwell).

—— Morris, M. G., Shaeil, J., Ward, L. K., Wells, D. A., and Wells, T. C. E., 1974, *Grassland ecology and wildlife management* (Chapman and Hall).

DUVIGNEAUD, P., and Denaeyer-de Smet, S., 1970, 'Biological cycling of minerals in temperate deciduous forests', in Reichle, D. E. (ed.), *Analysis of temperate forest ecosystems* (Ecological Studies 1, Chapman and Hall), 199–225.

DYNE, G. M. VAN (ed.), 1969, *The ecosystem concept in natural resource management* (Academic Press).

EDWARDS, A. M. C., 1973a, 'Dissolved load and tentative solute budget of some Norfolk catchments', *Journal of Hydrology,* **18,** 201–17.

—— 1973b, 'The variation of dissolved constituents with discharge in some Norfolk rivers', *Journal of Hydrology,* **18,** 219–42.

EDWARDS, C. A., Reichle, D. E., and Crossley, D. A., 1970, 'The role of soil invertebrates in turnover of organic matter and nutrients', in: Reichle, D. E. (ed.), *Analysis of temperate forest ecosystems* (Ecological Studies 1, Chapman and Hall), 147–72.

EDWARDS, N. T., and Sollins, P., 1973, 'Continuous measurement of carbon dioxide evolution from partitioned forest floor components', *Ecology,* **54,** 406–12.

EGNER, H., and Eriksson, E., 1955, 'Current data on chemical composition of air and precipitations', *Tellus,* **7,** 134–9.

ELGAWHARY, S. M., and Lindsay, W. L., 1972, 'Solubility of silica in soils', *Proceedings of the Soil Science Society of America,* **36,** 439–42.

EMERY, F. E. (ed.), 1969, *Systems thinking* (Penguin).

ENO, C. F., and Reuszer, H. W., 1950, 'Availability of potassium in certain minerals to *Aspergillus niger*', *Proceedings of the Soil Science Society of America,* **15,** 155.

ENVIRONMENTAL RESOURCES LIMITED, 1983, *Acid Rain: A review of the phenomenon in the EEC and Europe,* (Graham & Trotman, for the Commission of the European Communities).

ERIKSSON, E., 1952, 'The physico-chemical behaviour of nutrients in soils', *Journal of Soil Science,* **3,** 238-50.

—— 1960. 'The yearly cycle of chloride and sulphur in nature: meteorological, geochemical and pedological implications', I: *Tellus,* **11,** 335-40; II: *Tellus,* **12,** 63-109.

—— and Kuhnakasem, V., 1968, 'The chemistry of ground waters', in Eriksson, E., *et al. (eds.), Ground water problems* (Pergamon), 89-112.

FETH, J. H., Robertson, G., and Polzer, W., 1964, 'Sources of mineral constituents in water from granitic rocks, Sierra Nevada, California and Nevada', *United States Geological Survey, Water Supply Paper,* **1535-1.**

FISHER, D. W., Gambell, A. W., Likens G. E., and Bormann, F. H., 1968, 'Atmospheric contributions to water quality of streams in the Hubbard Brook Experimental Forest, New Hampshire', *Water Resources Research,* **4,** 1115-26.

FORTESCUE, J. A. C., and Marten, G. G., 1970, 'Micronutrients: forest ecology and systems analysis', in Reichle, D. E. (ed.), *Analysis of temperate forest ecosystems* (Ecological Studies 1. Chapman and Hall), 173-98.

FOWLER, D., Cape, J. N., and Leith, I. D., 1985, 'Acid inputs from the atmosphere in the United Kingdom', *Soil Use and Management,* **1,** 3-5.

FRIED, M., and Broeshart, H., 1967, *The soil-plant system* (Academic Press).

FRISSEL, M. J., and Reiniger, P., 1974, *Simulation of accumulation and leaching in soils* (Centre for Agricultural Publications and Documentation, Wageningen, Simulation Monographs).

FROMENT, A., 1972, 'Soil respiration in a mixed oak forest', *Oikos,* **23,** 273-7.

GALLOWAY, J. N., Norton, S. A., and Robbins Church, M., 1983, 'Freshwater acidification from atmospheric deposition of sulfuric acid: A conceptual model', *Environmental Science & Technology,* **17,** 541A-5A.

GAMBELL, A. W., and Fisher, D. W., 1966, 'Chemical composition of rainfall, eastern North Carolina and southeastern Virginia', *United States Geological Survey, Water Supply Paper,* **1535-K,** K1-K41.

GAMBLE, D. S., and Schnitzer, M., 1973, 'The chemistry of fulvic acid and its reaction with metal ions', in Singer, P. C. (ed.). *Trace metals and metal-organic interactions in natural waters* (Ann Arbor Science), 265-302.

GARDINER, W. C., 1969, *Rates and mechanisms of chemical reactions* (W. A. Benjamin).

GARDNER, W. R., and Brooks, R. H., 1957, 'A descriptive theory of leaching', *Soil Science,* **83,** 295-304.

GARRELS, R. M., and Christ, C. L., 1965, *Solutions, minerals and equilibria* (Harper and Row).

GARRETT, H. E., and Cox, G. S., 1973, 'Carbon dioxide evolution from the floor of an oak-hickory forest, *Proceedings of the Soil Science Society of America,* **37,** 641-4.

GERMANN, P. and Beven, K., 1981, 'Water flow in soil macropores. I: An experimental approach', *Journal of Soil Science,* **32,** 1-13.

GILBERT, O., and Bocock, K. L., 1960, 'Change in leaf litter when placed on the surface of soils with contrasting humus types', II: 'Changes in the nitrogen content of oak and ash leaf litter', *Journal of Soil Science,* **11,** 10-19.

GIMINGHAM, C. H., 1972, *The ecology of heathlands* (Chapman and Hall).

GLEN, R. R., 1967, 'Insecticides and environment', *Journal of Economic Entomology,* **60,** 398–405.

GOLD, V., 1956, *pH measurements: Their theory and practice* (Methuen).

GORHAM, E., 1955, 'On the acidity and salinity of rain', *Geochimica et Cosmochimica Acta,* **7,** 231–9.

—— 1957, 'The chemical composition of rain from Rosscahill in Co. Galway', *Irish Naturalists Journal,* **12,** 122–6.

—— 1958, 'The influence and importance of daily weather conditions in the supply of chloride, sulphate and other ions to fresh waters from atmospheric precipitation', *Philosophical Transactions of the Royal Society* (London), Ser. B, **241,** 147–78.

—— 1961, 'Factors influencing supply of major ions to inland waters with special reference to the atmosphere', *Bulletin of the Geological Society of America,* **72,** 795–840.

GORING, C. A. I., and Hamaker, J. W., 1972 *Organic chemicals in the soil environment,* vols. 1 and 2 (Marcel Dekker).

GOSZ, J. R., Likens, G. E., and Bormann, F. A., 1973, 'Nutrient release from decomposing leaf and branch litter in the Hubbard Brook Forest, New Hampshire', *Ecological Monographs,* **43,** 173–91.

GOUDIE, A. S., Cooke, R. W., and Evans, I., 1970. 'Experimental investigation of rock weathering by salts', *Area,* **4,** 42–8.

GOULD, R. F. (ed.), 1967, *Equilibrium concepts in natural water systems* (Advances in Chemistry Series 67, American Chemical Society).

—— (ed.) 1971, *Nonequilibrium systems in natural water chemistry* (Advances in Chemistry Series 106, American Chemical Society).

GRAY, P. H. H., and Wallace, R. H., 1957, 'Correlation between bacterial numbers and carbon dioxide in a field soil', *Canadian Journal of Microbiology,* **3,** 191.

GREENLAND, D. J. and Hayes, M. H. B., 1981, *The chemistry of soil processes* (Wiley).

GREGORY, K. J., and Walling, D. E. (eds.), 1974, *Fluvial processes in instrumented watersheds* (Institute of British Geographers, Special Publication, no. 6).

GRIME, J. P., and Hodgson, J., 1969, 'An investigation of the ecological significance of lime-chlorosis by means of large-scale comparative experiments', in Rorison, I. (ed.), *Ecological aspects of the mineral nutrition of plants* (Symposium of the British Ecological Society, no. 9, Blackwell).

GROBLER, J. H., and Pauli, F. W., 1964, 'Adsorption equilibria of humic substances', *South African Journal of Agricultural Science,* **7,** 187–92.

GRUBB, P. J., Green, H. E., and Merrifield, R. J. C., 1969, 'The ecology of chalk heath: its relevance to the calcicole–calcifuge and soil acidification problems', *Journal of Ecology,* **57,** 175–212.

—— and Suter, M. B., 1971, 'The mechanism of acidification of soil by Calluna and Ulex and the significance for conservation', in Duffey, E., and Watt, A. S., (eds.), *The scientific management of animal and plant communities for conservation* (Blackwell).

HAINES-YOUNG, R. and Petch, J. 1986, *Physical geography: Its nature and methods* (Harper & Row).

HALLSWORTH, E. G., and Crawford, D. V. (eds.), 1965, *Experimental pedology* (Butterworth).

HANDLEY, W. R. C., 1954, 'Mull and mor in relation to forest soils', *Forestry Commission Bulletin,* **23** (H.M.S.O.).

HARVEY, D. W., 1969, *Explanation in geography* (Edward Arnold).

HATTERI, T., 1973, *Microbial life in the soil* (Marcel Dekker).

HAWARTH, R. D., 1968, 'The nature of soil humic acid', *Biochemical Journal,* **109**, 950.

HEM, J. D., 1960, 'Restraints on dissolved ferrous iron imposed by bicarbonate, redox potential and pH', *United States Geological Survey, Water Supply Paper,* **1459-B**, 33–55.

—— 1970, 'Study and interpretation of the chemical characteristics of natural water', *United States Geological Survey, Water Supply Paper,* **1473**.

—— and Cropper, W. H., 1959, 'Survey of ferrous–ferric chemical equilibria and redox potentials', *United States Geological Survey, Water Supply Paper,* **1459-A**, 1–31.

HENDERSON, M. E. K., and Duff, R. B., 1963, 'The release of metallic and silicate ions from minerals, rocks and soils by fungal activity', *Journal of Soil Science,* **14**, 236–46.

HENDERSON, W., and Wilkins, C. W., 1975, 'The interaction of bush fires and vegetation', *Search, Journal of the Australian and New Zealand Association for the Advancement of Science,* **6**,(4), 130–3.

HERRERA, R., 1981. 'How human activities disturb the nutrient cycles of a tropical rain forest in Amazonia', *Ambio,* **10**, 109–14.

HORNUNG, M., 1985, 'Acidification of soils by trees and forests', *Soil Use and Management,* **1**, 24–8.

HORRILL, A. D., and Woodwell, G. M., 1973, 'Structure and cation content of a podzolic soil of Long Island, N Y, seven years after destruction of the vegetation by chronic gamma irradiation', *Ecology,* **54**, 439–44.

HOWARD, A. D., 1965, 'Geomorphological systems—equilibrium and dynamics', *American Journal of Science,* **263**, 302–12.

HUANG, W. H., and Keller, W. D., 1972, 'Organic acids as agents of chemical weathering of silicate minerals', *Nature, Physical Science,* **239**, 149–51.

—— and Kiang, W. C., 1972, 'Laboratory dissolution of plagioclase feldspars in water and organic acids at room temperature', *American Mineralogist,* **57**, 1849–59.

HURD, L. E., Mellinger, M. V., Wolf, L. L., and McNaughton, S. J., 1971, 'Stability and diversity at three trophic levels in terrestrial successional ecosystems', *Science* **173**, 1134–6.

IMESON, A. C., 1971, 'Heather burning and soil erosion on the North Yorkshire Moors', *Journal of Applied Ecology,* **8**, 537–42.

ISKANDER, I. K., and Syers, J. K., 1972, 'Metal-complex formation by lichen compounds', *Journal of Soil Science,* **23**, 255–65.

ISMAIL, F. T., 1970, 'Biotite weathering and clay formation in arid and humid regions, California', *Soil Science,* **109**, 257–61.

JACKSON, M. L., and Sherman, M. L., 1953, 'Chemical weathering of minerals in soils', *Advances in Agronomy,* **5**, 219.

JACKSON, R. M., and Raw, F., 1966, *Life in the soil* (Studies in Biology, no. 2, Edward Arnold).

JEFFRIES, R. L., Laycock, D., Stewart G. R., and Sims, A. P., 1969, 'The properties of mechanisms involved in the uptake and utilization of calcium and potassium by plants in relation to an understanding of plant distribution', in Rorison, I. (ed.), *Ecological aspects of the mineral nutrition of plants* (British Ecological Symposium 9, Blackwell), 281–307.

JOHNSON, N. M., 1971, 'Mineral equilibria in ecosystem geochemistry', *Ecology,* **52,** 529–31.

JONES, H. E., 1971, 'Comparative studies of plant growth and distribution in relation to waterlogging', I I: 'An experimental study of the relationship between transpiration and the uptake of iron in *Erica cinerea* and *E. tetralix, L., Journal of Ecology,* **59,** 167–78.

JONG, E. DE, and Schappert, H. J. V., 1972, 'Calculation of soil respiration and activity from CO_2 profiles in the soil', *Soil Science,* **113,** 328–83.

JORDAN, C. F. and Herrera, R., 1981, 'Tropical rain forests: Are nutrients really critical?', *American Naturalist,* **117,** 167–80.

JORDAN, K. F., 1985, *Nutrient cycling in tropical forest ecosystems: principles and applications in management and conservation* (Wiley).

JUNGE, C. E., and Gustafson, P. E., 1956, 'Precipitation sampling for chemical analysis', *Bulletin of the American Meteorological Society,* **37,** 244–5.

JURINAK, J. J., and Griffin, R. A., 1972, 'Nitrate ion adsorption by calcium carbonate', *Soil Science,* **113,** 130–5.

KELLER, W. D., 1957, *The principles of chemical weathering* (Lucas).

—— and Frederickson, A. F., 1952, 'Role of plants and colloidal acids in the mechanism of weathering', *American Journal of Science,* **250,** 594–608.

KEMPER, W. D., Sills, I. D., and Aylmore, L. A. G., 1970, 'Separation of adsorbed cation species as water flows through clays', *Proceedings of the Soil Science Society of America,* **34,** 946–58.

KERN, O. M., 1960, 'The hydration of carbon dioxide', *Journal of Chemical Education,* **37,** 14–23.

KERPEN, W., and Scharpenseel, H. W., 1967, 'Movements of ions and colloids in undisturbed soil and parent material columns', in *Isotope and radiation techniques in soil physics and irrigation studies* (Proceedings of Symposium, FAO/IAEA, Istanbul), 213–26.

KIRKBY, M. J., 1969, 'Erosion by water on hillslopes', in Chorley, R. J. (ed.), *Water, earth and man* (Methuen), 229–38.

KITTREDGE, J., 1948, (repub. 1973), *Forest influences: The effect of woody vegetation on climate, water and soil* (Dover Publications).

KLINE, J. R., 1973, 'Mathematical simulation of soil–plant relationships and soil genesis', *Soil Science,* **115,** 240–9.

KONONOVA, M. M., 1966, *Soil organic matter* (Pergamon).

KOZLOWSKI, T. T., and Ahlgren, C. E., 1974, *Fire and ecosystems* (Academic Press).

KURTZ, L. T., and Melsted, S. W., 1973, 'Movements of chemicals in soils by water', *Soil Science,* **115,** 231–9.

LÅG, J., 1968, 'Relationships between the chemical composition of the precipitation and the contents of exchangeable ions in the humus layer of natural soils', *Acta Agriculturae Scandinavica,* **18,** 148–52.

—— 1976, 'Influence of soils on fresh water', in Skreslet, S., Leinebø, R., Matthews, J. B. L., and Sakshaug, E. (eds.), *Fresh Water on the Sea,* Proceedings from a symposium on the influence of fresh water outflow on biological processes, in fjords and coastal waters, 22–5 April 1974, Geilo, Norway. (Association of Norwegian Oceanographers, Oslo).

LEAKE, B. E., *et al.*, 1969, 'The chemical analysis of rock powders by automatic X-Ray fluorescence', *Chemical Geology,* **5**, 7–86.

LEE, G. F., and Hoadley, A. W., 1967, 'Biological activity in relation to chemical equilibrium composition of natural waters', in Gould, R. F. (ed.), *Equilibrium concepts in natural water systems,* (Advances in Chemistry Series 67, American Chemical Society), 319–38.

LEOPOLD, L. B., Wolman, M. G., and Miller, J. P., 1964, *Fluvial processes in geomorphology* (W. H. Freeman and Co.).

LEWIN, J., Cryer, R., and Harrison, D. I., 1974, 'Sources for sediments and solutes in mid-Wales', in Gregory, K. J., and Walling, D. E. (eds.), *Fluvial processes in instrumented watersheds* (Institute of British Geographers, Special Publication, no. 6), 73–85.

LIDDLE, M. J., 1975, 'A theoretical relationship between the primary productivity of vegetation and its ability to tolerate trampling', *Biological Conservation,* **8**, 251–6.

LIKENS, G. E., Bormann, F. H., Pierce, R. S., and Fisher, D. W., 1970a, 'Nutrient-hydrologic cycle interactions in small forested watershed-ecosystems', in d'Andigni de Asis, L. (ed.), *Productivity of forest ecosystems* (Proceedings of the Brussels Symposium, 1969, UNESCO Publications, Paris), pp. 553–63.

—— Johnson, N. M., Fisher, D. W., and Pierce, R. S., 1970b, 'Effects of forest cutting and herbicide treatment on nutrient budgets in the Hubbard Brook watershed ecosystem', *Ecological Monographs,* **40**, 23–47.

—— —— Pierce, R. S., Eaton, J. S., and Johnson, N. M., 1977, *Biogeochemistry of a forested ecosystem* (Springer-Verlag).

LINDSAY, W. L., 1979, *Chemical equilibria in soils* (Wiley-Interscience).

LINDSTROM, F. T., and Boersma, L., 1971, 'A theory on the mass transport of previously distributed chemicals in a water saturated sorbing porous medium', *Soil Science,* **111**, 192–9.

—— 1973, 'A theory on the mass transport of previously distributed chemicals in a water-saturated sorbing porous medium', III: 'Exact solution for first-order kinetic solution', *Soil Science,* **115**, 5–10.

—— and Stockard, D., 1971, 'A theory on the mass transport of previously distributed chemicals in a water-saturated sorbing porous medium, II: 'Isothermal cases', *Soil Science,* **112**, 291–300.

LIVINGSTONE, D. A., 1963, 'Chemical composition of rivers and lakes', in Fleischer, M. (ed.), *Data of geochemistry, United States Geological Survey, Professional Paper,* **440–G**, G1–G64.

LOUGHNAN, F. C., 1969, *Chemical weathering of the silicate minerals* (Elsevier).

LOVERING, T. S., and Engel, C., 1967, 'Translocation of silica and other elements from rock into Equisetum and three grasses', *United States Geological Survey, Professional Paper,* **594–B**, B1–B16.

LUKASHEV, K. I., 1970, *Lithology and geochemistry of the weathering crust* (Israel Programme for Scientific Translation).

LUNDEGARDH, H., 1951, *Leaf analysis* (Hilger and Watts).

MADGWICK, H. A. I., and Ovington, J. D., 1959, 'The chemical composition of precipitation in adjacent forest and open plots', *Forestry,* **32**, 14–22.

MAJOR, J., 1970, 'Essay review of Rodin and Bazilevich: the illusive mineral equilibrium', *Ecology,* **51**, 160–3.

MALTBY, E., 1975, 'Numbers of soil micro-organisms as ecological indicators of changes resulting from moorland reclamation on Exmoor', *Journal of Biography,* **2**, 117–36.

—— and Crabtree, K., 1976, 'Soil organic matter and peat accumulation on Exmoor: A contemporary and palaeoenvironmental evaluation', *Transactions of the Institute of British Geographers,* NS1, 259–78.

MANSKAYA, S. M., and Drozdova, T. V., 1968, *Geochemistry of organic substances* (International Series of Monographs in Earth Sciences 28, Pergamon).

—— 1969, 'Organic matter as a factor in rock weathering', in Khitarov, N. I. (ed). *Problems of geochemistry* (Israel Programme for Scientific Translation).

MARSHALL, C. E., Chowdury, M. Y., and Upchurch, W. J., 1973, 'Lysimetric and chemical investigations of pedological changes', pt. 2: 'Equilibriation of profile samples with aqueous solutions', *Soil Science,* **116**, 336–58.

MARTEL, A. E., and Calvin, M., 1952, *Chemistry of the metal chelate compounds* (Prentice-Hall).

MASON, J., 1986, 'Acid precipitation' in Conway, G., (ed.), *The assessment of environmental problems,* Imperial College, Centre for Environmental Technology, ch. 2, 5–10.

MASSEY, H. F., and Jackson, M. L., 1952, 'Selective erosion of soil fertility constituents', *Proceedings of the Soil Science Society of America,* **16**, 353–6.

McCOLL, J. G., 1969, 'Ion transport in a forest soil: Models and mechanisms' (Univ. of Washington Ph.D. thesis).

McGINNIS, J. T., Golley, F. E., Clements, R. G., Child, G. I., and Duever, M. J., 1969, 'Elemental and hydrologic budgets of the Panamanian tropical moist forest', *Bioscience,* **19**, 697–700.

McLAREN, A. D., and Peterson, G. H., 1967, *Soil biochemistry,* vol. 1 (Marcel Dekker).

—— and Skujins, J., 1971, *Soil Biochemistry,* vol. 2 (Marcel Dekker).

MERCADO, A., and Billings, G. K., 1975, 'The kinetics of mineral dissolution in carbonate aquifers as a tool for hydrological investigation', I: 'Concentration–time relationships', *Journal of Hydrology,* **24**, 303–31.

MEYBECK, M., 1976, 'Total mineral dissolved transport by world major rivers', *Hydrological Science Bulletin,* **21**, 265–84.

MILLER, H. G., 1985, 'The possible role of forests in streamwater acidification', *Soil Use and Management* **1**, 28–9.

MITCHELL, J., 1973, 'Mobilisation of phosphorous by *Pteridium aquilinum*', *Plant and Soil,* **38**, 489–91.

MOHR, E. C. J., Baren, P. A. van, and Schuylenborgh, J., 1972, *Tropical Soils* (Mouton, The Hague).

MORGAN, J. J., 1967, 'Applications and limitations of chemical thermodynamics in natural water systems', in Gould, R. F. (ed.), *Equilibrium concepts in natural water systems* (Advances in Chemistry Series 67, American Chemical Society), 1–29.

MURRAY, A. N., and Love, W. W., 1929, 'Action of organic acids upon limestone', *Bulletin of the American Association of Petroleum Geologists,* **13**, 1467–75.

NERNST, W., 1904, 'Theorie der Reaktionsgeschwindigkeit in heterogenen Systemem', Zeistchrift für physikalische Chemie, **47**, 52–5.

NEWBOULD, P., 1969, 'The absorption of nutrients by plants from different zones in the soil', in Rorison, I. H. (ed.), *Ecological aspects of the mineral nutrition of plants* (Symposium of British Ecological Society 9, Blackwell).

NEWMAN, L., Likens, G. E., and Bormann, F. H., 1975, 'Acidity in rainwater: Has an explanation been presented?', *Science,* **188,** 957.

NIHLGARD, B., 1972, 'Plant biomass, primary production and distribution of chemical elements in a beech and a planted spruce forest in south Sweden', *Oikos,* **23,** 69–81.

NILSSON, I., 1972, 'Accumulation of metals in spruce needles and needle litter', *Oikos,* **23,** 132–6.

NYE, P. H., 1972, 'The measurement and mechanism of ion diffusion in soils', V I I I: 'Theory for the propagation of changes of pH in soils', *Journal of Soil Science,* **23,** 82–92.

—— and Tinker, P. B. H., 1975, *Solute transport and plant growth* (Blackwell).

OBORN, E. T., and Hem, J. D., 1961, 'Microbial factors in the solution and transport of iron', *United States Geological Survey, Water Supply Paper, 1459–H.*

ODUM, E. P., 1962, 'Relationship between structure and function in the ecosystem', *Japanese Journal of Ecology,* **12,** 108–18, quoted from Watts, D., 1971, *Principles of Biogeography* (McGraw-Hill), 41.

OLLIER, C. D., 1975, *Weathering* (Longman).

ONG, H. L., and Bisque, R. E., 1968, 'Coagulation of humic colloids by metal ions', *Soil Science,* **106,** 220–4.

ONG, H. L., Swanson, V. E., and Bisque, R. E., 1970, 'Natural organic acids as agents of chemical weathering', *United States Geological Survey, Professional Paper,* **700–C,** 130–7.

O'RIORDAN, T., 1971, *Perspectives on resource management* (Pion).

OVINGTON, J. D., 1962, 'Quantitative ecology and the woodland ecosystem concept', *Advances in Ecological Research,* **1,** 103–83.

OWENS, M., 1970, 'Nutrient balance in rivers', *Water Treatment and Examination,* **19,** 239–47.

PATTEN, B. E., (ed.), 1971, *Systems analysis and simulation in ecology,* vol. 1 (Academic Press).

—— (ed.), 1972, *Systems analysis and simulation in ecology,* vol. 2 (Academic Press).

—— (ed.), 1975, *Systems analysis and simulation in ecology,* vol.3 (Academic Press).

PEARCE, A. J., 1976, 'Magnitude and frequency of erosion by Hortonian overland flow', *Journal of Geology,* **84,** 65–80.

PERRIN, R. M. S., 1965, 'The use of drainage water analyses in soil studies', in Hallsworth, E. G., and Crawford, D. V., (eds.), *Experimental pedology* (Butterworth), 73–92.

PICKETT, S. T. A. and White, P. S., 1985, *The ecology of natural disturbance and patch dynamics* (Academic Press).

PITTY, A. F., 1966, 'An approach to the study of karst water', *Occasional Papers in Geography,* **5**(Univ. of Hull).

PLATT. J. R., 1964, 'Strong inference', *Science,* **146** (3642), 347–53.

POLZER, W. L., 1967, 'Geochemical controls of solubility of aqueous silica', in Faust,

S. D., and Hunter, J. V. (eds.), *Principles and applications of water chemistry* (Wiley), 505–19.

PREISNITZ, K., 1972, 'Methods of isolating and quantifying solution factors in the laboratory', *Transactions of the Cave Research Group of Great Britain,* **14,** 153–8.

RACKHAM, O., 1986, *The history of the countryside* (J. M., Dent & Sons).

RAMSEY, J. A., 1971, *A guide to thermodynamics* (Chapman and Hall).

RAPP, A., 1974, 'A review of desertization in Africa—Water, vegetation and man', *Secretariat for International Ecology, Stockholm, Sweden,* Report No. 1.

REDFIELD, A. C., 1958, 'The biological control of chemical factors in the environment', *American Naturalist,* **46,** 205–21.

REICHLE, D. E. (ed.), 1970, *Analysis of temperate forest ecosystems* (Ecological Studies 1, Chapman and Hall).

REINERS, W. A., 1968, 'Carbon dioxide evolution from the floor of three Minnesota forests', *Ecology,* **49,** 471–83.

RENNIE, P. J., 1955, 'The uptake of nutrients by mature forest growth', *Plant and Soil,* **7,** 49.

RICHARDS, B. N., 1974, *Introduction to the soil ecosystem* (Longman).

RODIN, L. E., and Bazilevich, N. J., 1967, *Production and mineral cycling in terrestrial vegetation,* trans G. E. Fogg (Oliver and Boyd).

RORISON, I. H. (ed.), 1969, *Ecological aspects of the mineral nutrition of plants* (Symposium of British Ecological Society 9, Blackwell).

ROSE, C. W., 1962, 'Some effects of rainfall, radiant drying and soil factors on infiltration under rainfall into soils', *Journal of Soil Science,* **13,** 286–97.

RUNGE, E. C. A., 1973, 'Soil development sequences and energy models', *Soil Science,* **115,** 183–93.

RUSSEL, E. W., 1973, *Soil conditions and plant growth* (Longman).

SCHNITZER, M., 1959, 'Interaction of iron with rainfall leachates', *Journal Soil Science,* **10,** 300–8.

—— and Desjardins, J. G., 1962, 'Molecular and equivalent weights of the organic matter of a podzol', *Proceedings of the Soil Science Society of America,* **26,** 362–5.

—— and Khan. S. U., 1972, *Humic substances in the environment* (Marcel Dekker).

—— and Skinner, I. M, 1963, 'Organo-metallic interactions in soils', I: 'Reactions between a number of metal ions and the organic matter of a podzol Bh horizon', *Soil Science,* **96,** 86–93.

SCHUMM, S. A., and Lichty, R. W., 1965, 'Time, space and causality in geomorphology', *American Journal of Science,* **363,** 110–19.

SCHWARZENBACH, G., and Flaschka, H., 1969, *Complexiometric titrations* (Methuen).

SCRIVNER, C. L., Baker, J. C., and Brees, D. R., 1973, 'Combined daily climatic data and dilute solution chemistry in studies of soil profile formation', *Soil Science,* **115,** 213–23.

SHAFFER, P. W. and Galloway, J. N., 1982, 'Acid precipitation: the impact on two headwater streams in Shenandoah National Park, Virginia', *International Symposium on Hydrometeorology,* American Water Resources Association, 43–53.

SICCAMA, T. G., Bormann, F. H., and Likens, G. E., 1970, 'The Hubbard Brook ecosystem study: Productivity, nutrients and phytosociology of the herbaceous layer', *Ecological Monographs,* **40,** 389-402.

SINGER, P. C. (ed.), 1973, *Trace metals and metal-organic interactions in natural water systems* (Ann Arbor Science).

SLOBODKIN, L. B., 1962, 'Energy in animal ecology', in Cragg, J. B. (ed.), *Advances in Ecological Research,* vol. 1 (Academic Press), 69–101.

SMALL, E., 1972, 'Photosynthetic rates in relation to nitrogen recycling as adaptation to nutrient deficiency in peat bog plants', *Canadian Journal of Botany,* **50,** 2227–33.

SMITH, D. I., and Mead, D. G., 1962, 'The solubility of limestones with special reference to Mendip', *Proceedings of the University of Bristol Speleological Society,* **10,** 119–38.

—— and Newson, M. D., 1974, 'The dynamics of solutional and mechanical erosion in limestone catchments on the Mendip Hills, Somerset', in Gregory, K. J., and Walling, D. E. (eds.), *Fluvial processes in instrumental watershed* (Institute of British Geographers, Special Publication, no. 6).

SMITH, L. P., 1971, 'The problem of prediction', *Agricultural Meteorology,* **9,** 1–2.

SMITH, R. T., 1984, 'Soils in ecosystems', Ch. 7 in Taylor, J. A., (ed.), *Themes in biogeography* (Croom Helm).

SOPPER, W. E., and Lull, H. W. (eds.), 1967, *International symposium on forest hydrology* (Pergamon).

STARK, N. and C. F. Jordan, 1978. 'Nutrient retention by the root mat of an Amazonian rain forest', *Ecology,* **59,** 434–7.

STEENBJERG, J., 1954, 'The weathering of minerals as indicated by plants', *Journal of Soil Science,* **5,** 205–13.

STEINBECK, J., 1960, *The log from the Sea of Cortez* (Pan Books).

STEVENSON, C. M., 1968, 'An analysis of the chemical composition of rain water and air over the British Isles and Eire for the years 1959–1964', *Quarterly Journal of the Royal Meteorological Society,* **94,** 56–70.

STODDART, D. R., 1969, 'Climatic geomorphology: A review and re-assessment', *Progress in Geography,* **1,** 159–222.

STONE, E. L., and Kszystyniak, R., 1977, 'Conservation of potassium in the *Pinus resinosa* ecosystem', *Science,* **198,** 192–4.

STUMM, W., and Morgan, J. J., 1970, *Aquatic chemistry: An introduction emphasising chemical equilibria in natural waters* (Wiley).

SUTCLIFFE, J. F., and Baker, D. A., 1974, *Plants and mineral salts* (Studies in Biology, no. 48, Edward Arnold).

SWAIN, F. M., 1970, *Non-marine organic geochemistry* (Cambridge University Press).

SWANK, W. T., 1984, 'Atmospheric contributions to forest nutrient cycling', *Water Resources Bulletin,* **20,** 313–21.

—— 1986, 'Biological control of solute losses from forest ecosystems', in Trudgill, S. T. (ed.), *Solute processes* (Wiley).

—— and Henderson, G. S., 1976, 'Atmospheric input of some cations and anions to forest ecosystems in North Carolina and Tennessee', *Water Resources Research,* **12,** 541–6.

Sweeting, M. M., 1972, *Karst Landforms* (Macmillan).

Symposium (Society of Chemical Industry), 1973, 'Nutrient losses from soil by leaching and drainage', Soc. Chem. Ind. Report, Oct. 1972, *Journal of the Science of Food and Agriculture,* **24,** 479–83.

Tabatabai, M. Ali, 1985, 'Effect of acid rain on soils', *CRC Critical Reviews in Environmental Control,* **15,** 65–110.

Tallis, J. H., 1964, 'Studies of southern Pennine peats', II: 'The pattern of erosion', *Journal of Ecology,* **52,** 323–11; III: 'The behaviour of sphagnum', ibid. **52,** 345–53.

—— 1965, 'Studies of southern Pennine peats', IV: 'Evidence of recent erosion', *Journal of Ecology,* **53,** 509–20.

Tamm, C. O., 1951, 'The removal of plant nutrients from tree crowns by rain', *Physiologia Planta,* **4,** 1848.

Tamm, E., and Krzysch, G., 1963, 'The effect of soil temperature and moisture on carbon dioxide production in a loamy sand', Zeitschrift für Aiker-und Pflanzenbau, **117,** 359–78.

Thomas, A. S., 1963, 'Further changes in vegetation since the advent of myxomatosis', *Journal of Ecology,* **5,** 151–83.

Thomas, G. W. and Philips, R. E., 1979, 'Consequences of water flow through macropores', *Journal of Environmental Quality,* **8,** 149–52.

Thomas, W. A., 1969, 'Accumulation and cycling of calcium by dogwood trees', *Ecological Monographs,* **39,** 101–20.

Thorp, J., 1967, 'Effects of certain animals that live in soils', in Drew, J. V. (ed.), *Selected papers in soil formation and classification* (Soil Science Society of America, Special Publication, 1), 191–208.

Tolgyesi, G., Csapody, L., and Bencze, L., 1968, 'Analysis of the ash components of ligneous and herbaceous plants grown on acid primary rocks and calcareous parent material', *Agrokémia és talajian,* **17,** 225–36.

Trudgill, S. T., 1976, 'The erosion of limestones under soil and the long term stability of soil–vegetation systems on limestone', *Earth Surface Processes,* **1,** 31–41.

—— 1983, *Weathering and Erosion,* (Hutchinson).

—— (ed.), 1986, *Solute processes* (Wiley).

—— Pickles, A. M., and Smettem, K. R. J., 1983, 'Soil water residence time and solute uptake, II: Dye tracing and preferential flow predictions', *Journal of Hydrology,* **62,** 279–85.

Twidale, C. R., Bourne, J. A., and Smith, D. M., 1974, 'Reinforcement and stabilisation mechanisms on landform development', *Revue de geomorphologie dynamique,* XXIII[c] année 3.

Upchurch, W. J., Chowdhury, M. U., and Marshall, C. E., 1973, 'Lysimetric and chemical investigation of pedological changes', I: 'Lysimeters and their drainage waters', *Soil Science,* **116,** 266–81.

Verry, E. S., 1975, 'Streamflow chemistry and nutrient yields from upland-peatland watersheds in Minnesota', *Ecology,* **56,** 1149–57.

Vimmerstedt, J. P., and Finney, J. H., 1973, 'Impact of earthworm introduction on litter burial and nutrient distribution in Ohio strip-mine spoil banks', *Proceedings of the Soil Science Society of America,* **37,** 388–91.

VOIGT, K., 1980, 'Acid precipitation and soil buffering capacity', in Drabløs, D., and Tollan, A. (eds.), *Ecological impact of acid precipitation,* Proceedings International Conference, Sandefjord, Norway, 11–14 March 1980. SNS F Project, Box 16, 1432, Ås-NLH, Norway.

WALLWORK, J. A., 1970, *Ecology of soil animals* (McGraw-Hill).

WARING, R. H. and Schlesinger, W. H., 1985, *Forest ecosystems: Concepts and management* (Academic Press).

WATT, K. E. F. (ed.), 1966, *Systems analysis in ecology* (Academic Press).

—— 1968, *Ecology and resource management* (McGraw-Hill).

WEBB, A. H., 1980, 'The effect of chemical weathering on surface waters', in Drabløs, D., and Tollan, A., *Ecological impact of acid precipitation,* Proceedings of an International Conference, Sandefjord, Norway, 11–14 March 1980. SNSF, Project, Box 16, 1432, Ås-NLH, Norway.

WEYMAN, D. R., 1970, 'Throughflow on hillslopes and its relation to the stream hydrograph', *Bulletin of the International Association of Scientific Hydrology,* **15**, (2), 25–32.

—— 1975, *Runoff processes and streamflow modelling* (Oxford University Press).

WHITE, E., Starkey, R. S., and Saunders, M. J., 1971, 'An assessment of the relative importance of several chemical sources to the waters of a small upland catchment', *Journal of Applied Ecology,* **8**, 743–9.

WHITE, E. J., and Turner, F., 1970, 'A method of estimating income of nutrients in a catch of airborne particles by a woodland canopy', *Journal of Applied Ecology,* **7**, 441–61.

WHITHEAD, H. C., and Feth, J. H., 'Chemical composition of rain, dry fallout and bulk precipitation, California, 1957–1959', *Journal of Geophysical Research,* **9**, 3319–33.

WHITTAKER, R. H., Bormann, F. H., Likens, G. E., and Siccama, T. G., 1974, 'The Hubbard Brook ecosystem study: Forest biomass and production', *Ecological Monographs,* **44**, 233–54.

WIKLANDER, L., 1964, 'Cation and anion exchange phenomena', in Bear, F. E. (ed.), *Chemistry of the soil* (American Chemical Society, Monograph 16, Van Nostrand-Reinhold). ch. 4.

—— and Andersson, A., 1972, 'The replacing efficiency of hydrogen ions in relation to base saturation and pH', *Geoderma,* **7**, 159–65.

WILLIAMS, R. J. B., 1970, 'The chemical composition of water from land drains at Saxmundham and Woburn and the influence of rainfall upon nutrient losses', *Report of Rothamsted Experimental Station,* pt. 2, 36–37.

—— and Cooke, G. W., 1971, 'Results of the Rotation I experiment at Saxmundham, 1964–1969', *Report of Rothamsted Experimental Station,* pt. 2., 68–97.

WILLIAMS, R. J. P., 1953, 'Metal ions in biological systems', *Biological Review,* **28**, 381–415.

WILLIAMS, S. T., and Gray, T. R. G., 1974, 'Decomposition of litter on the soil surface', in Dickinson, C. H. and Pugh, G. J. F. (eds.), *Biology of plant litter decomposition,* vol. 2 (Academic Press), 611–32.

WIT, C. T. DE, and Keulen, H. van, 1972, *Simulation of transport processes in soils* (Centre for Agricultural Publications and Documentation, Wageningen, Simulation Monographs).

WOODWELL, G. M., and Whittaker, R. H., 1967, 'Primary production and the cation budget of the Brookhaven Forest', in *Symposium on primary productivity and mineral cycling in natural ecosystems,* Paper presented to American Association for the Advancement of Science, Annual Meeting New York (Maine University Press), 151–66.

WRIGHT, T. W., 1955, 'Profile development in the sand dunes of Culbin Forest, Morayshire', I: 'Physical properties', *Journal of Soil Science,* **6**, 270–83.

—— 1956, 'Profile development in the sand dunes of Culbin Forest, Morayshire', II: 'Chemical properties'. *Journal of Soil Science,* **7**, 33–42.

YAALON, D. H., and Lomas, J., 1976, 'Factors controlling the supply and chemical composition of aerosols in a near shore and coastal environment', *Agricultural Meteorology,* **7**, 370–81.

ZEMAN, L. J., and Slaymaker, H. O., 1975, 'Hydrochemical analysis of discriminate variable runoff source areas in an alpine basin', *Arctic and Alpine Research,* **7**, 341–51.

ZVEREV, V. P., 1968, 'Role of atmospheric precipitation in the rotation of chemical elements between the atmosphere and lithosphere', *Doklady Akademii Nauk SSSR,* **181**, 195–8.

Author Index

Author Index

Subject Index